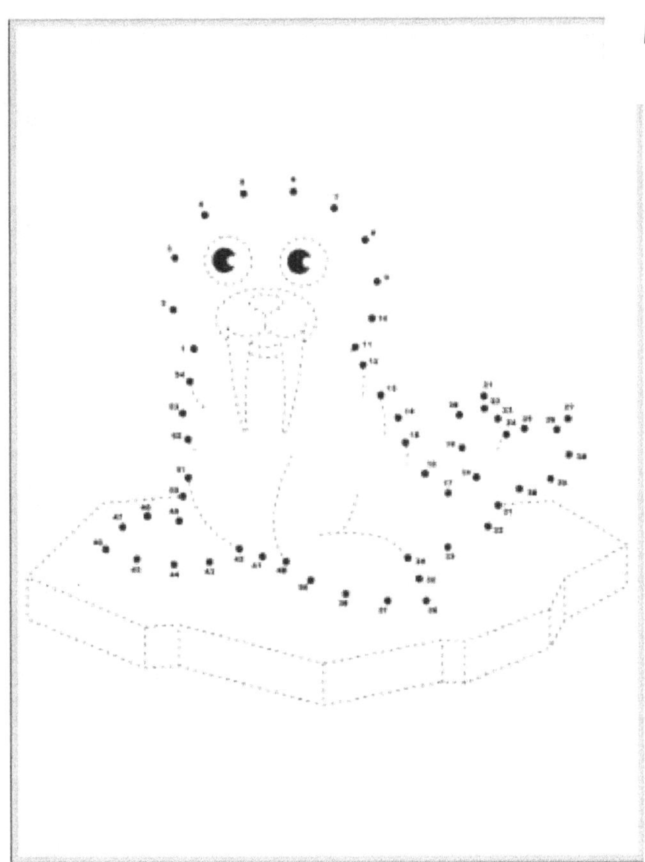

Thise Books Belongs To..

COLOR TEST PAGE

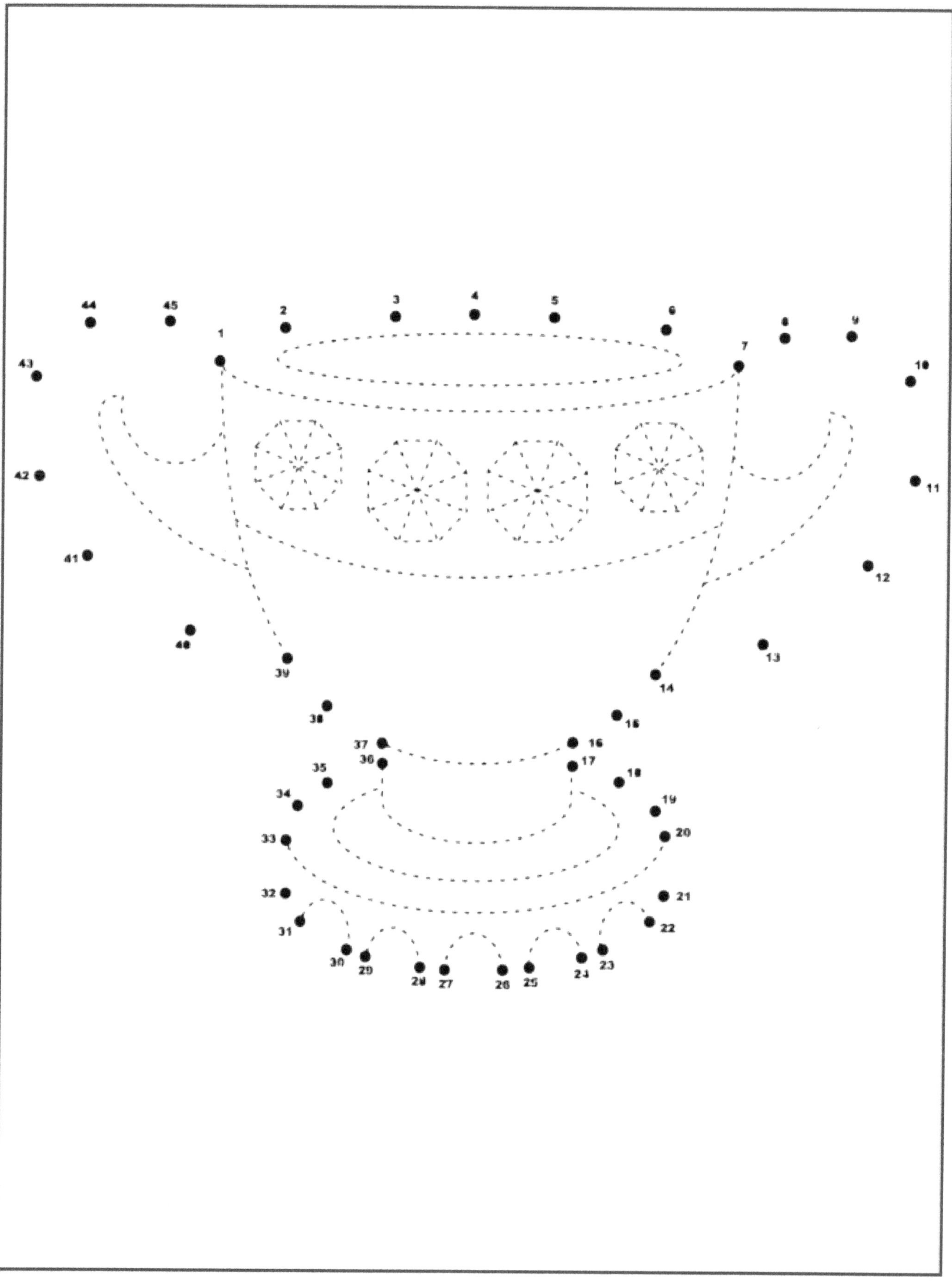

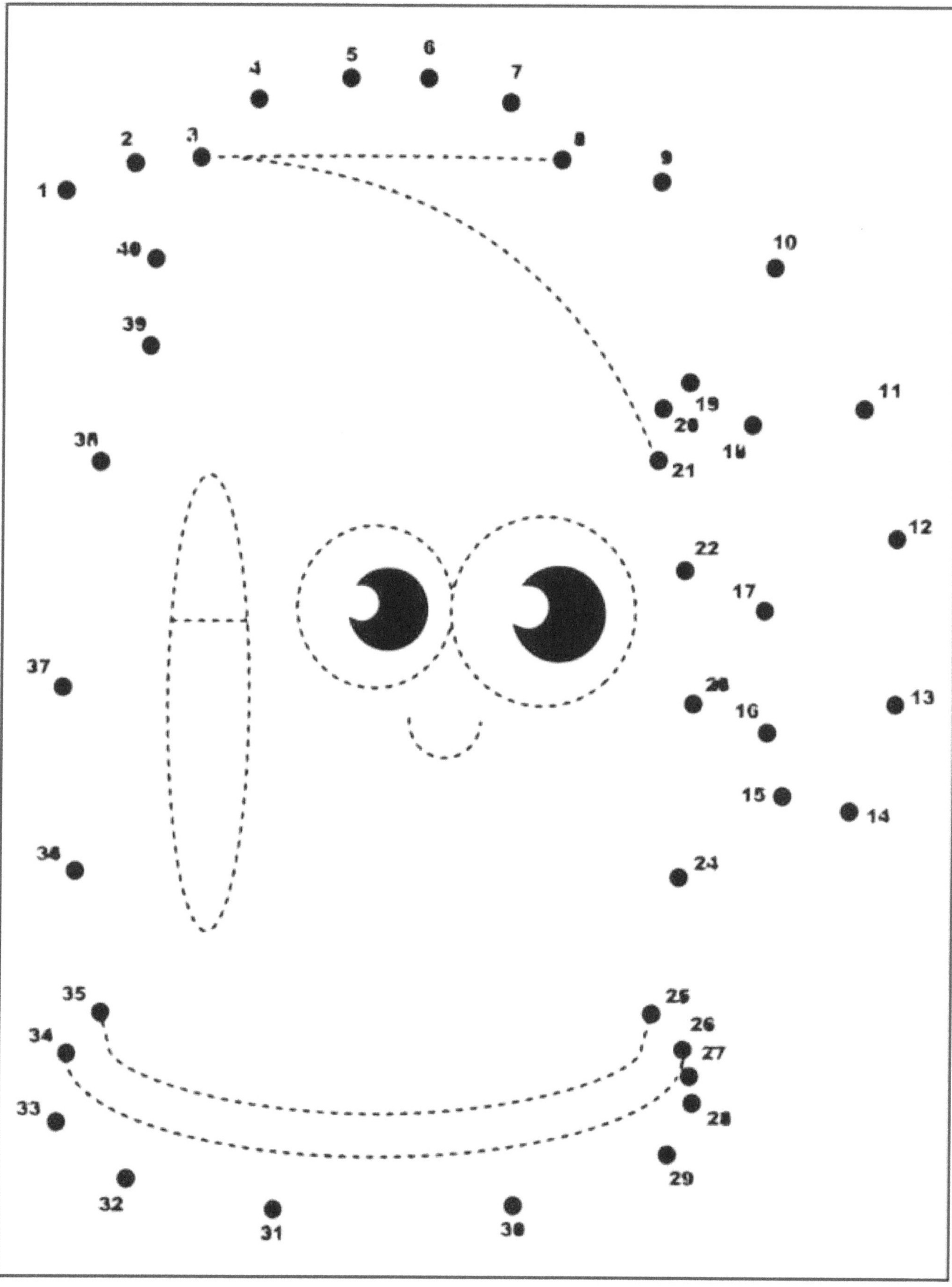

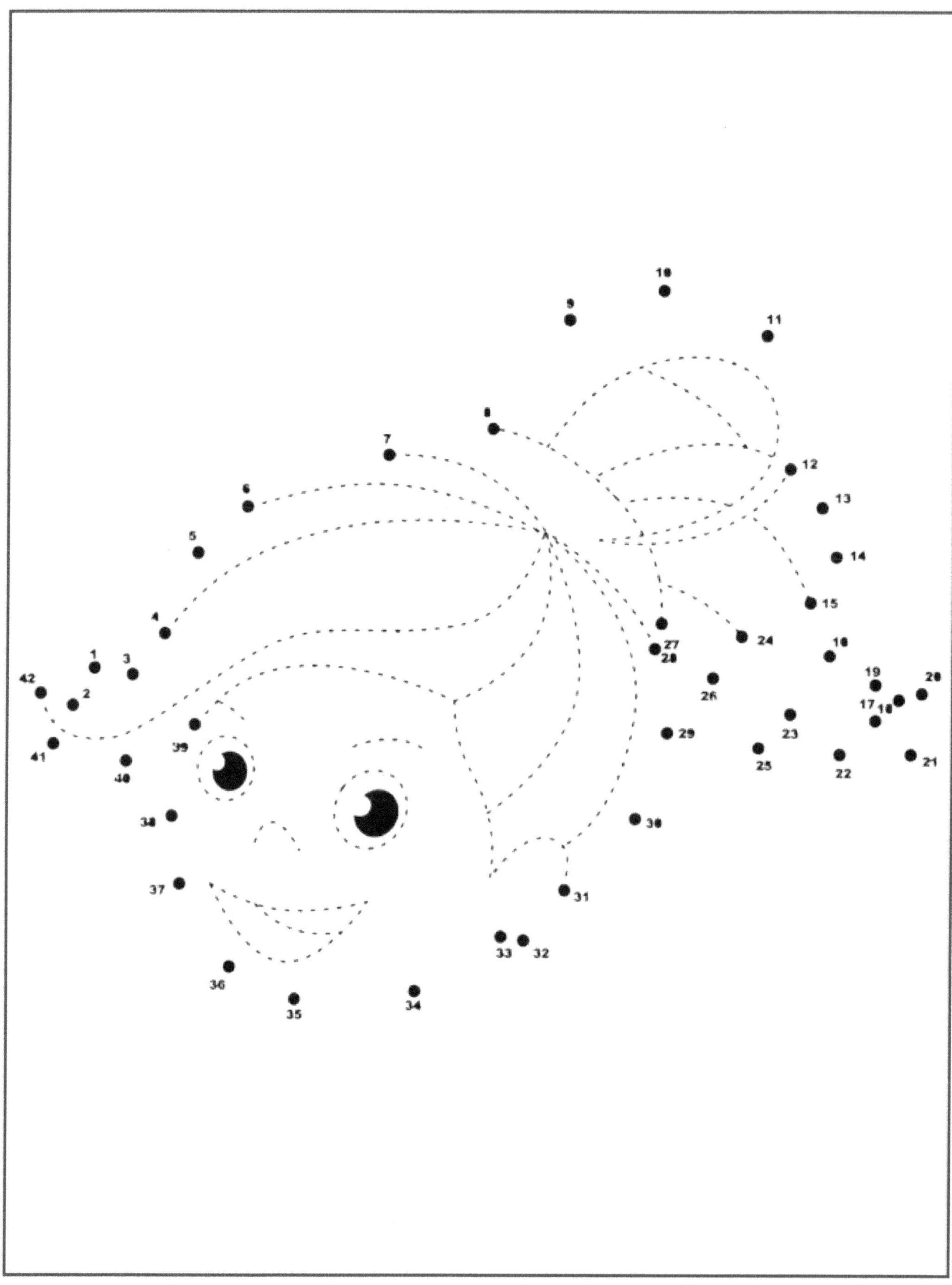

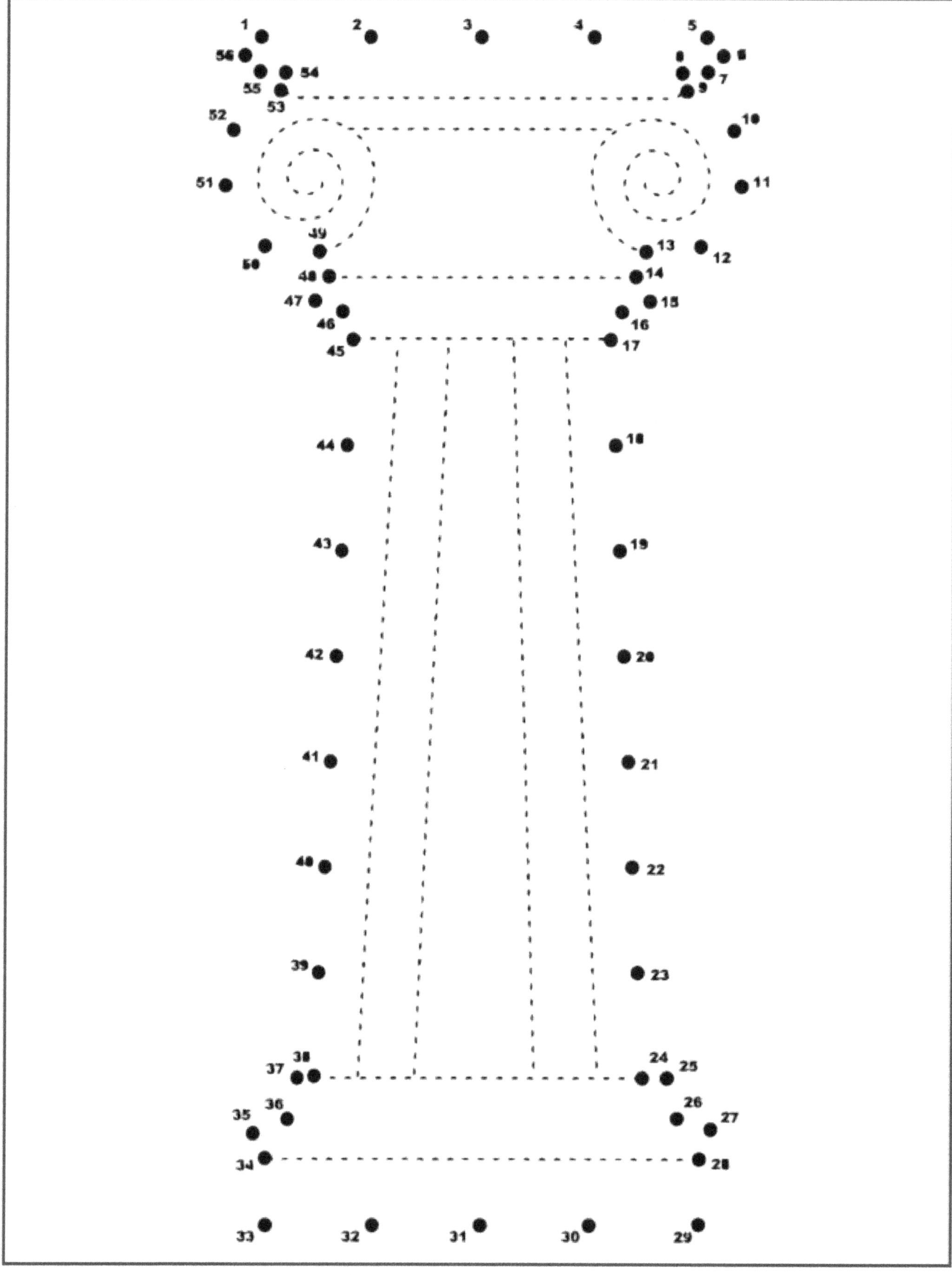

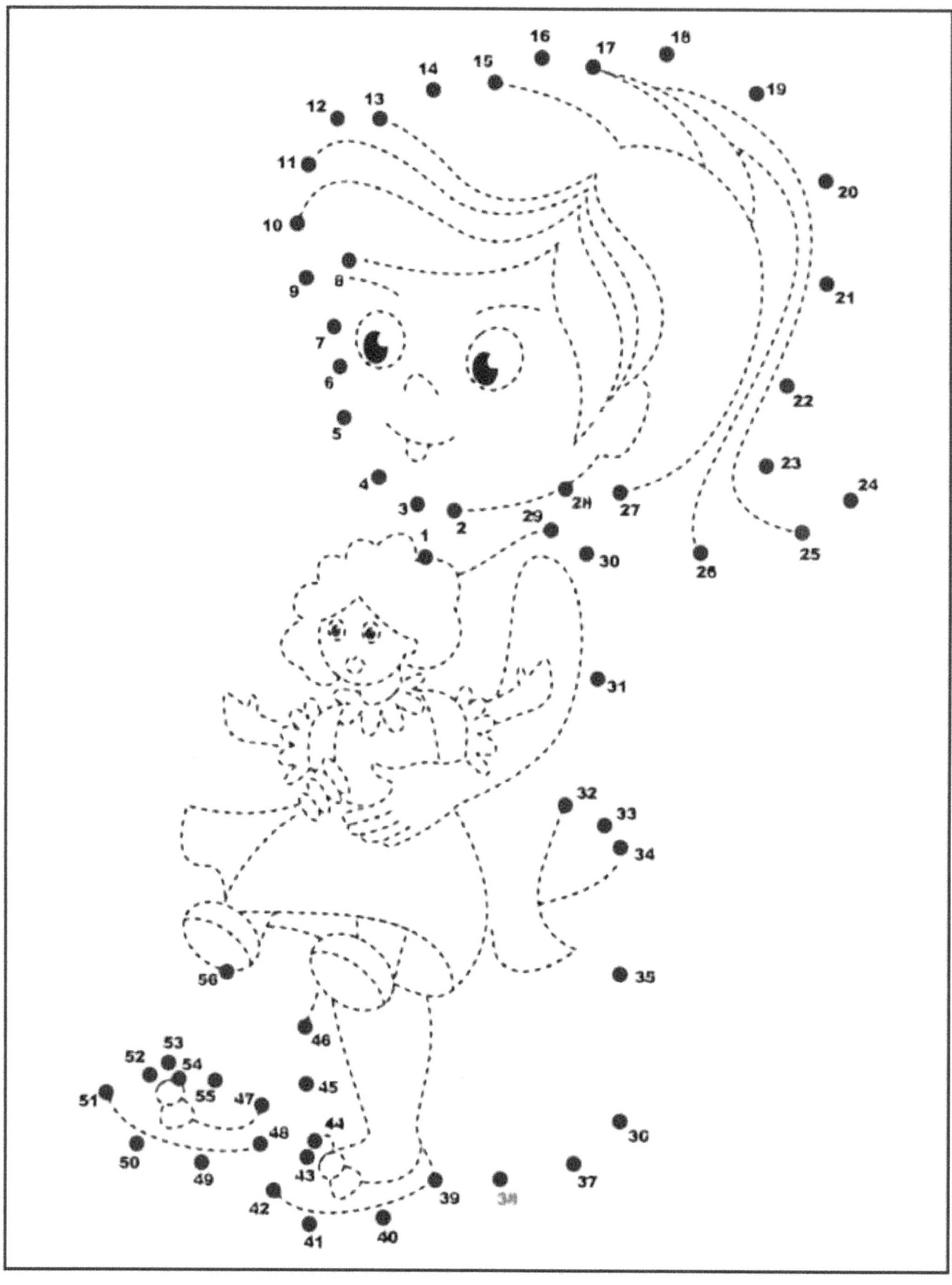

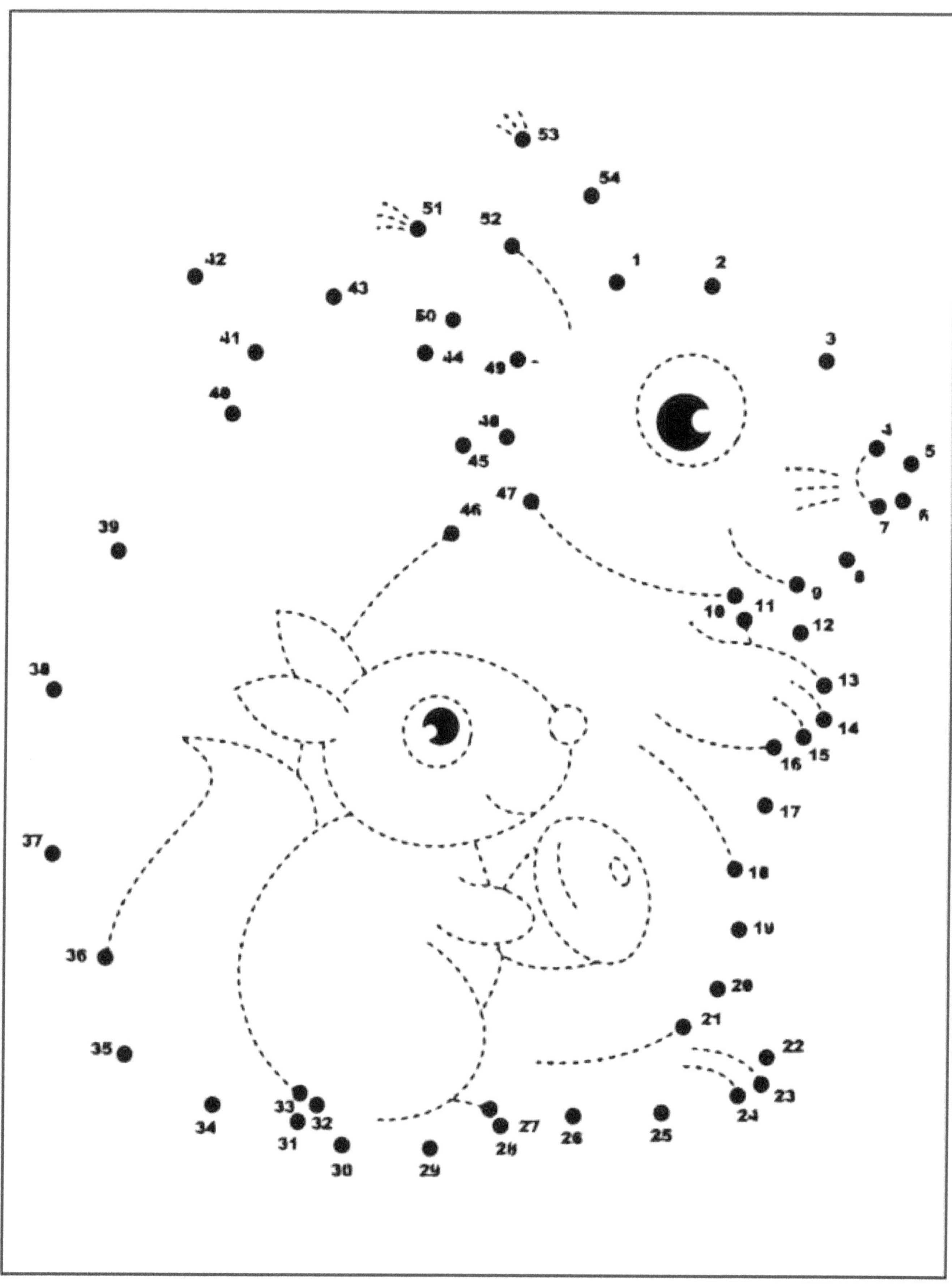

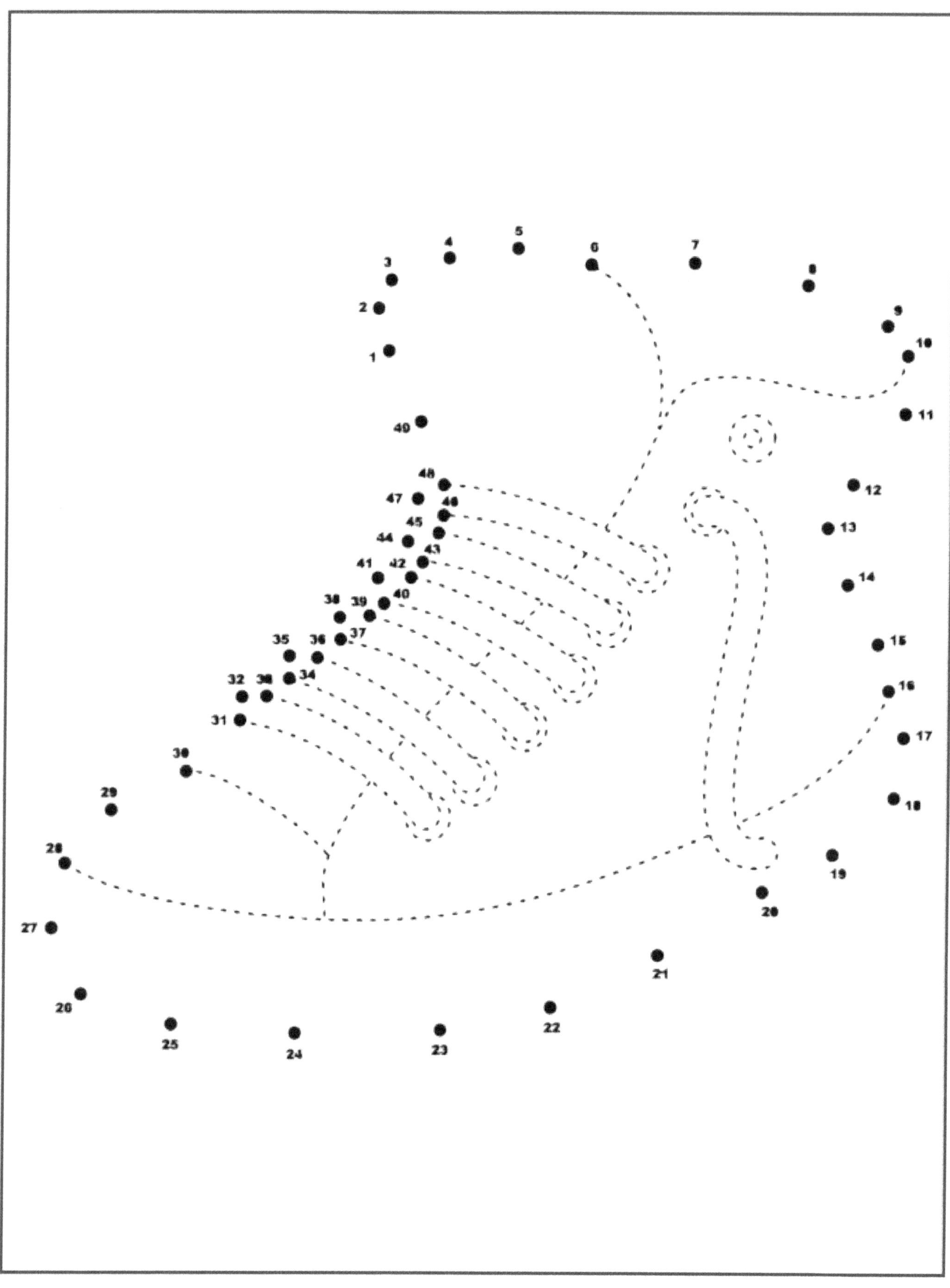

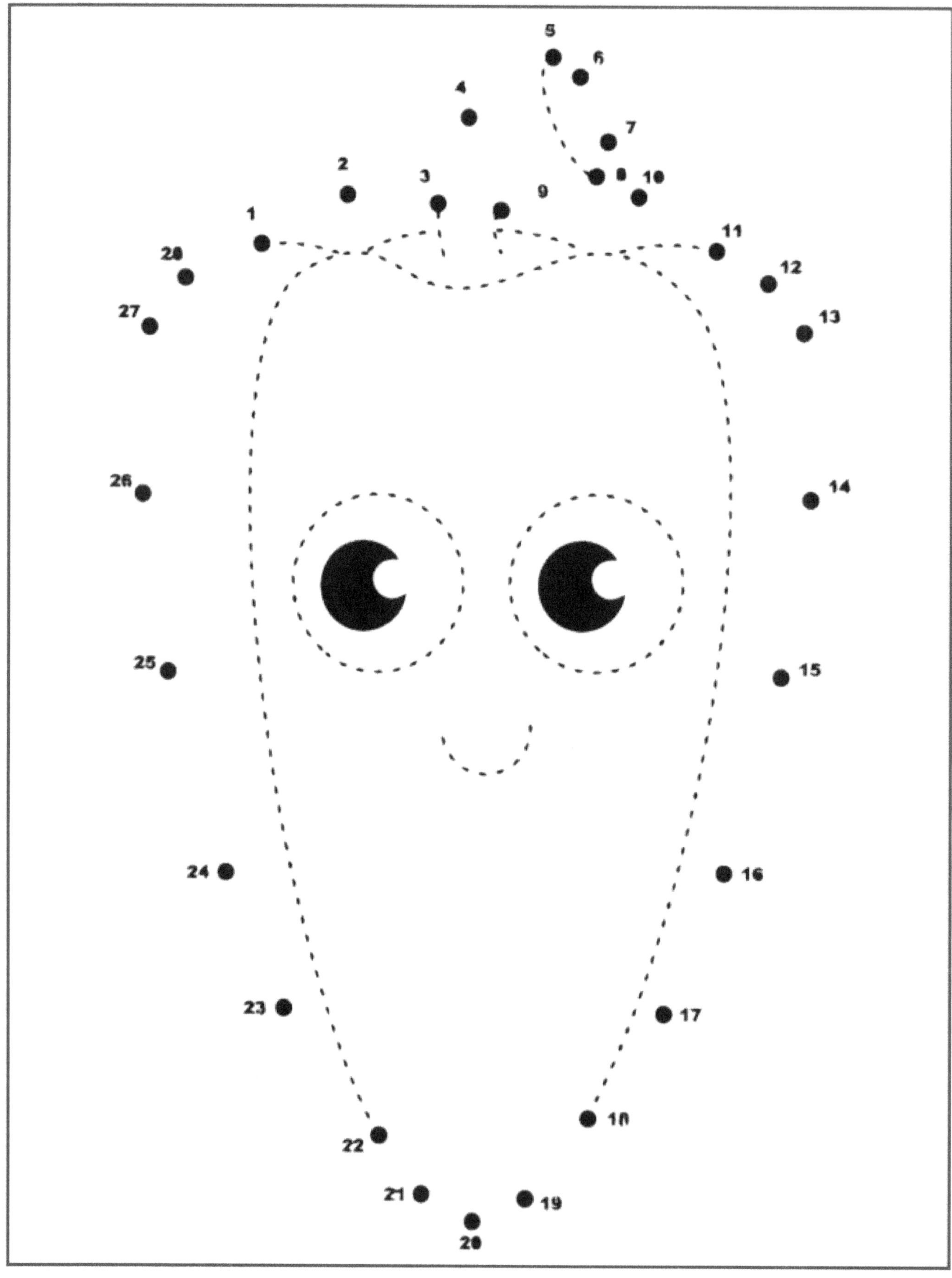

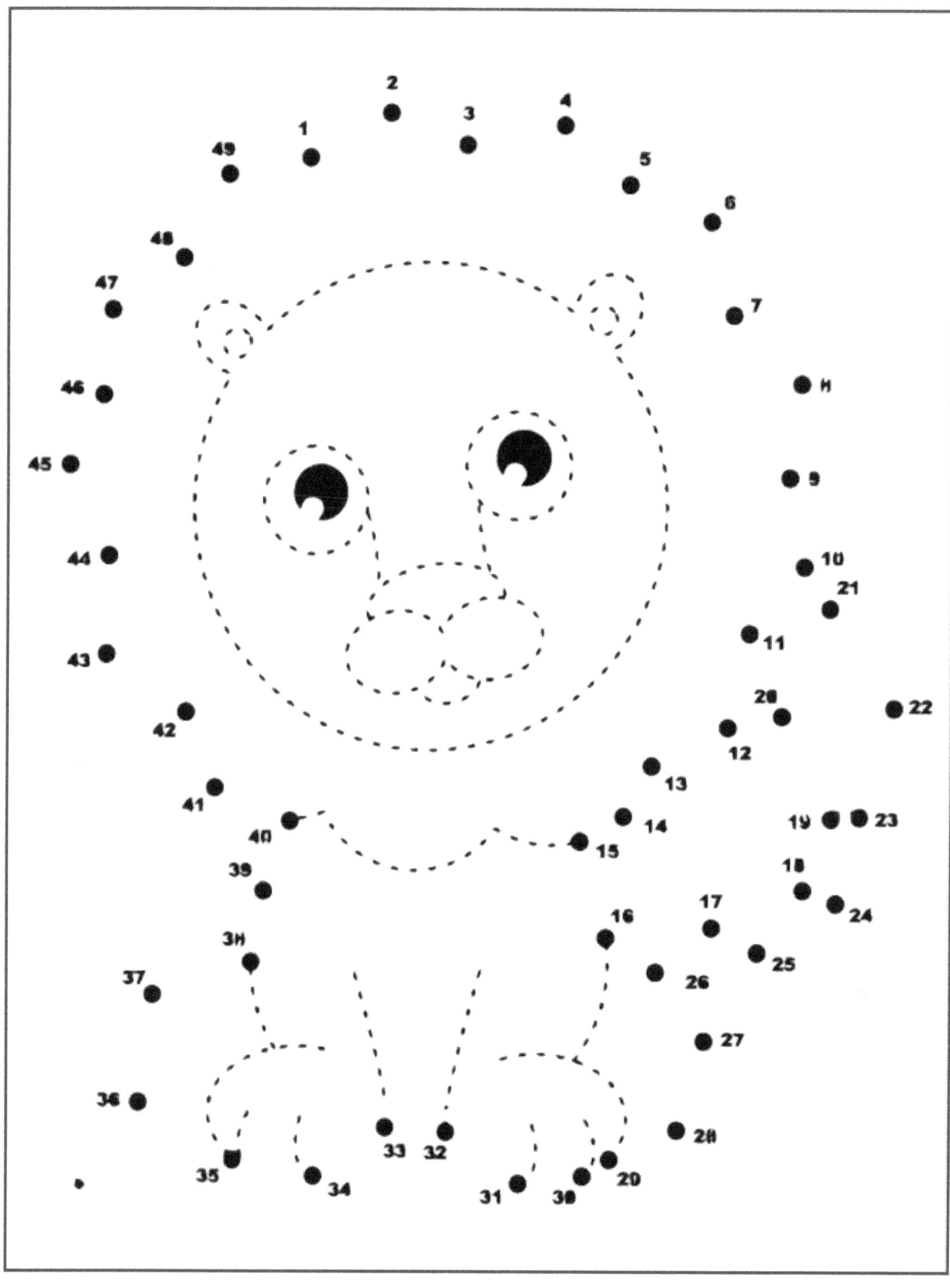

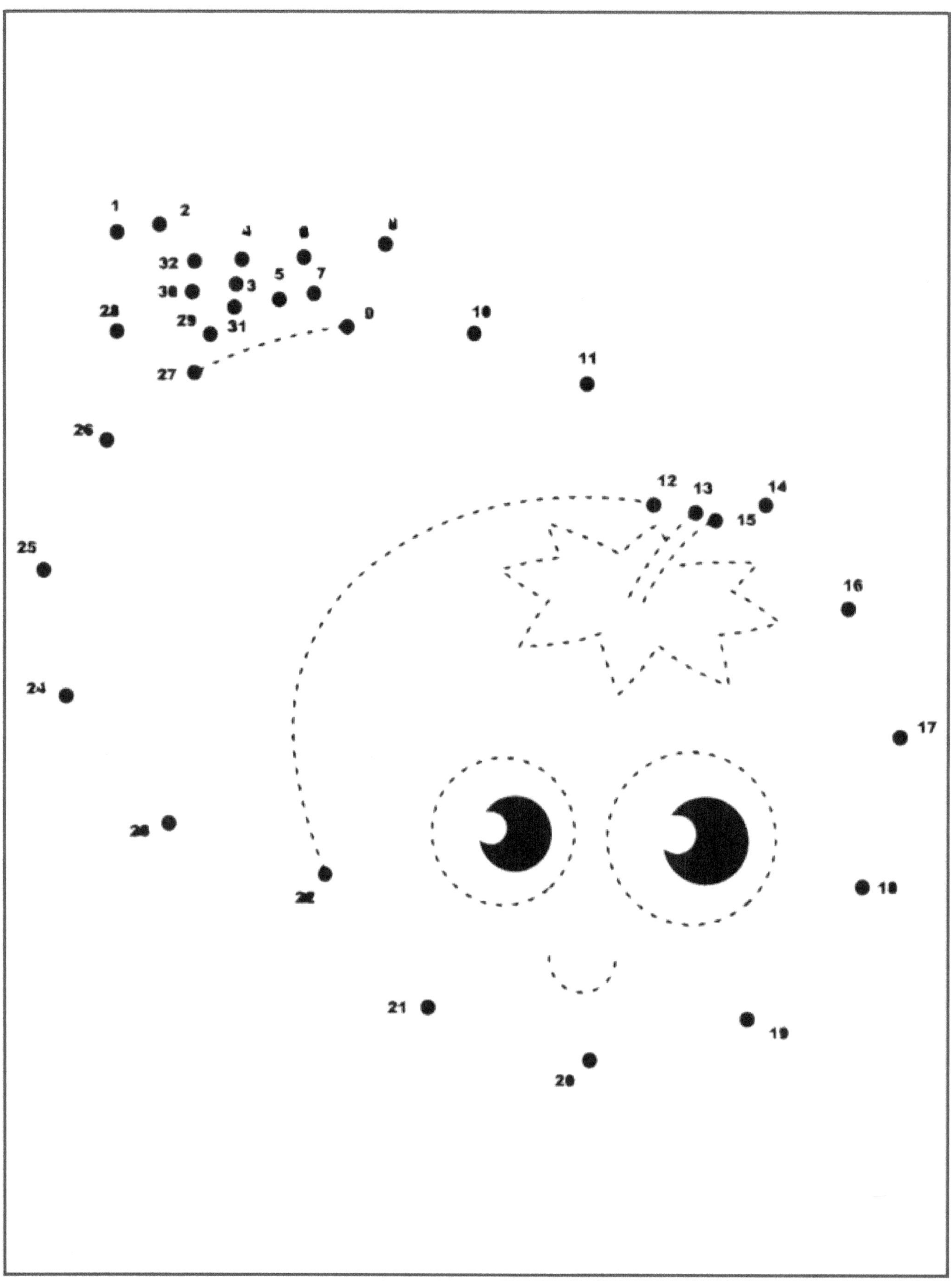

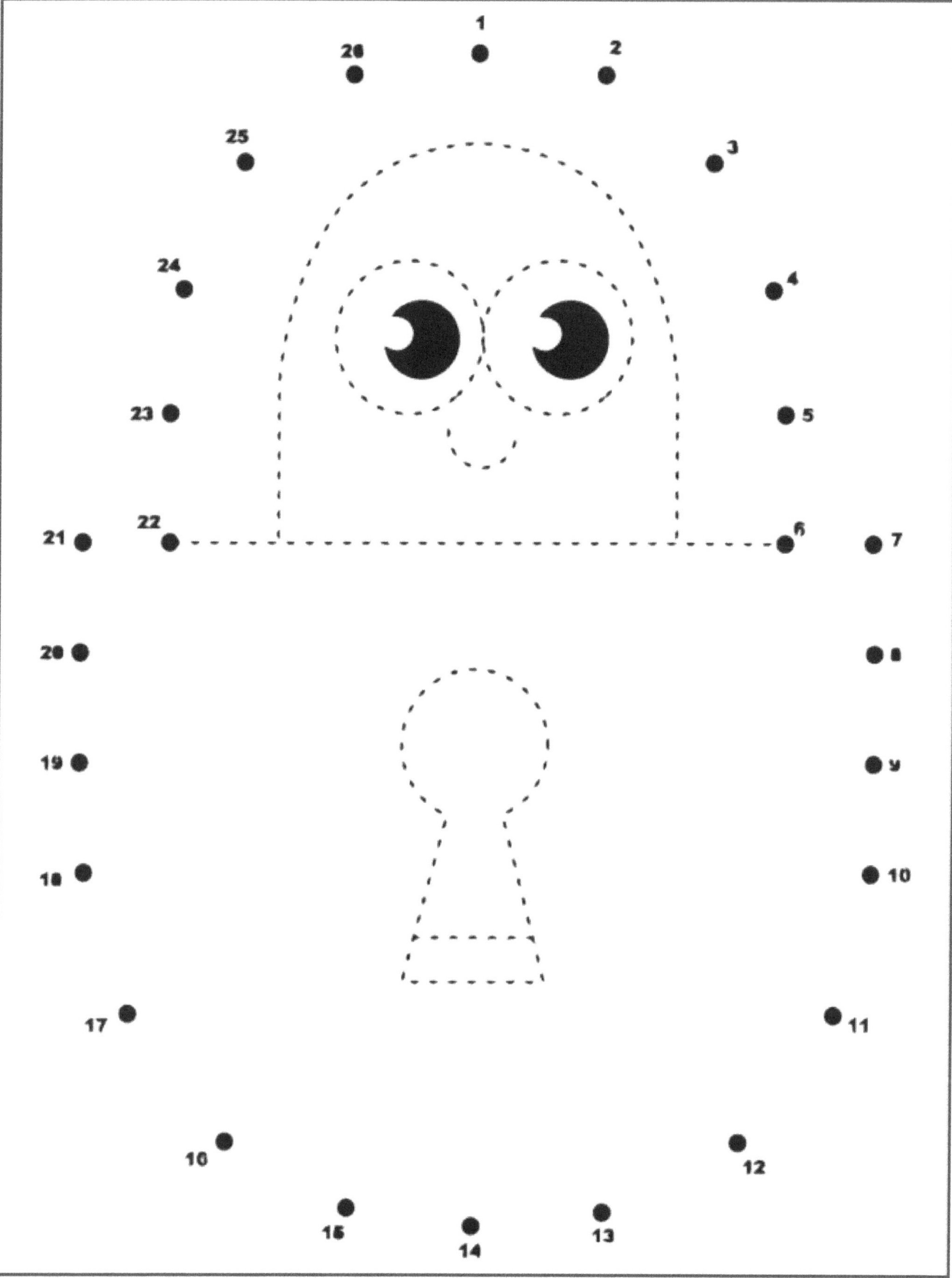

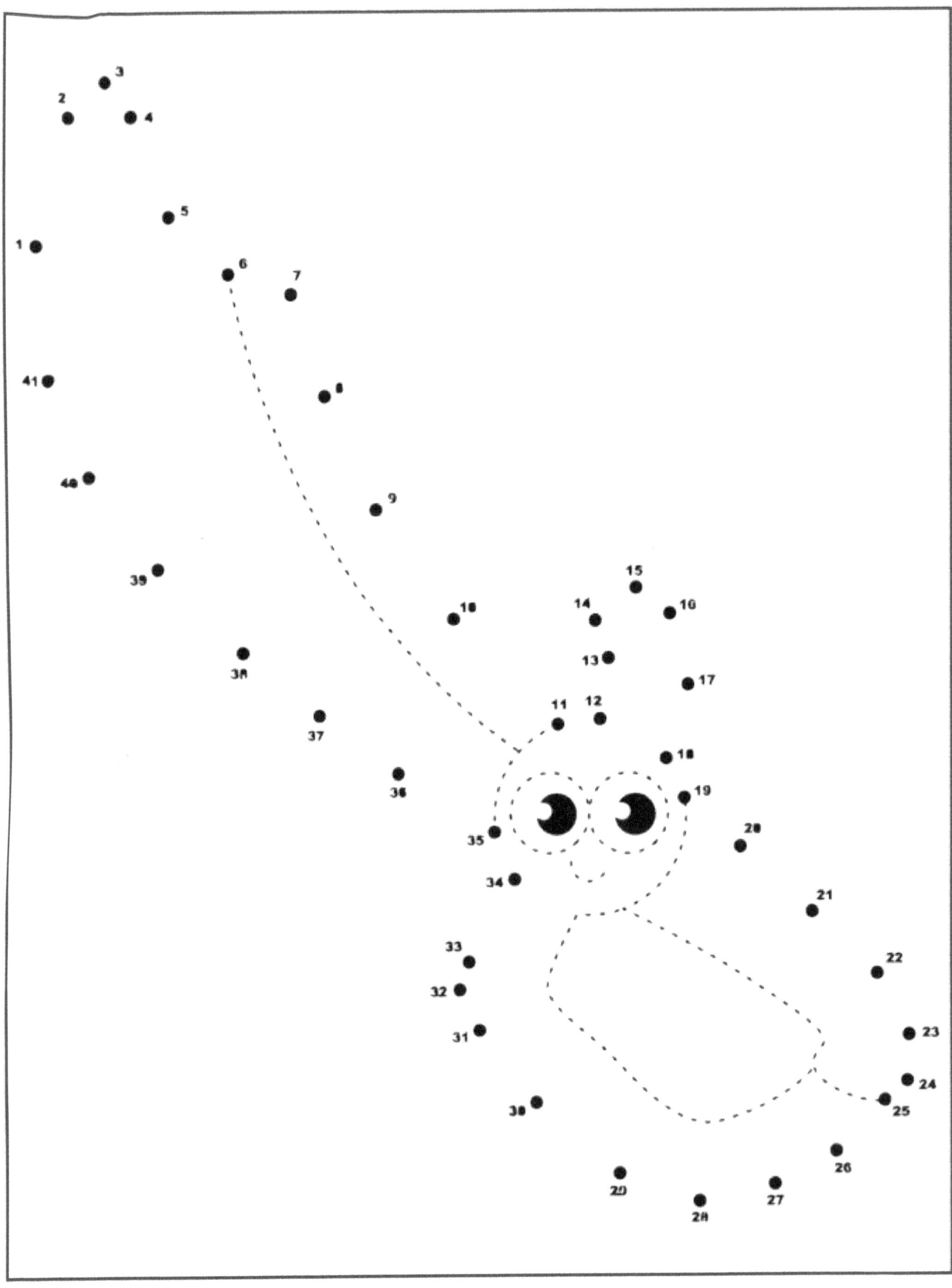

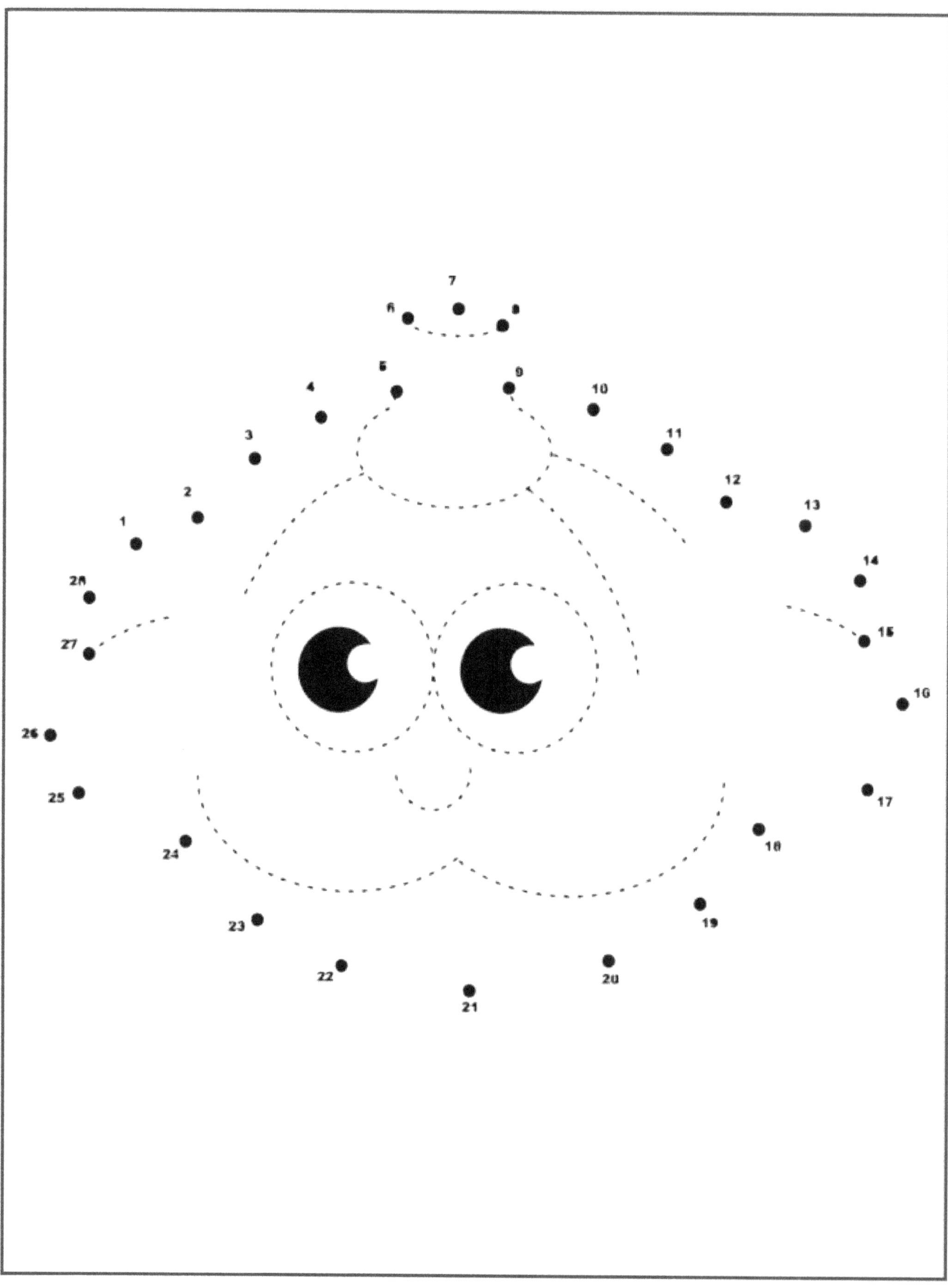

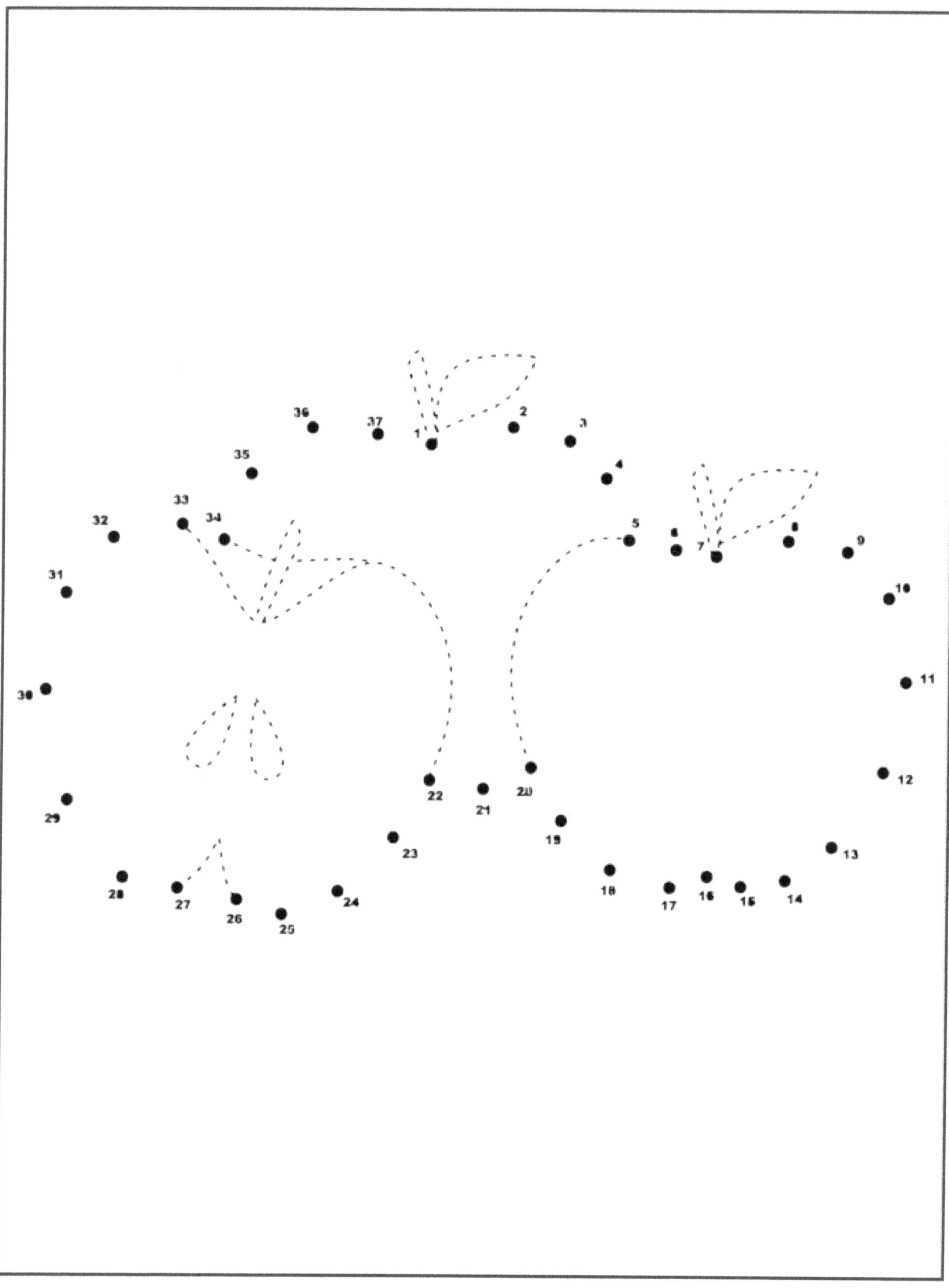

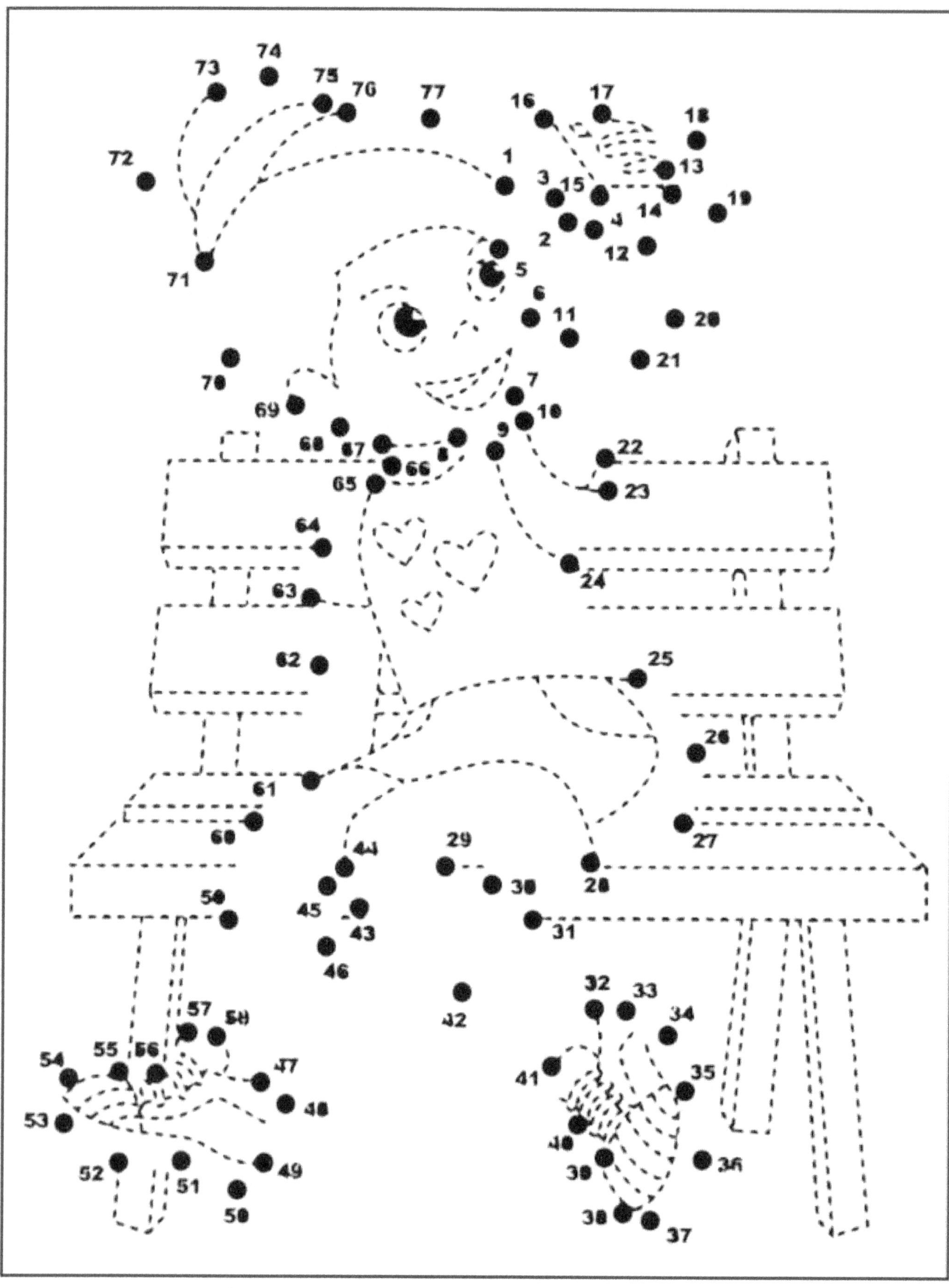

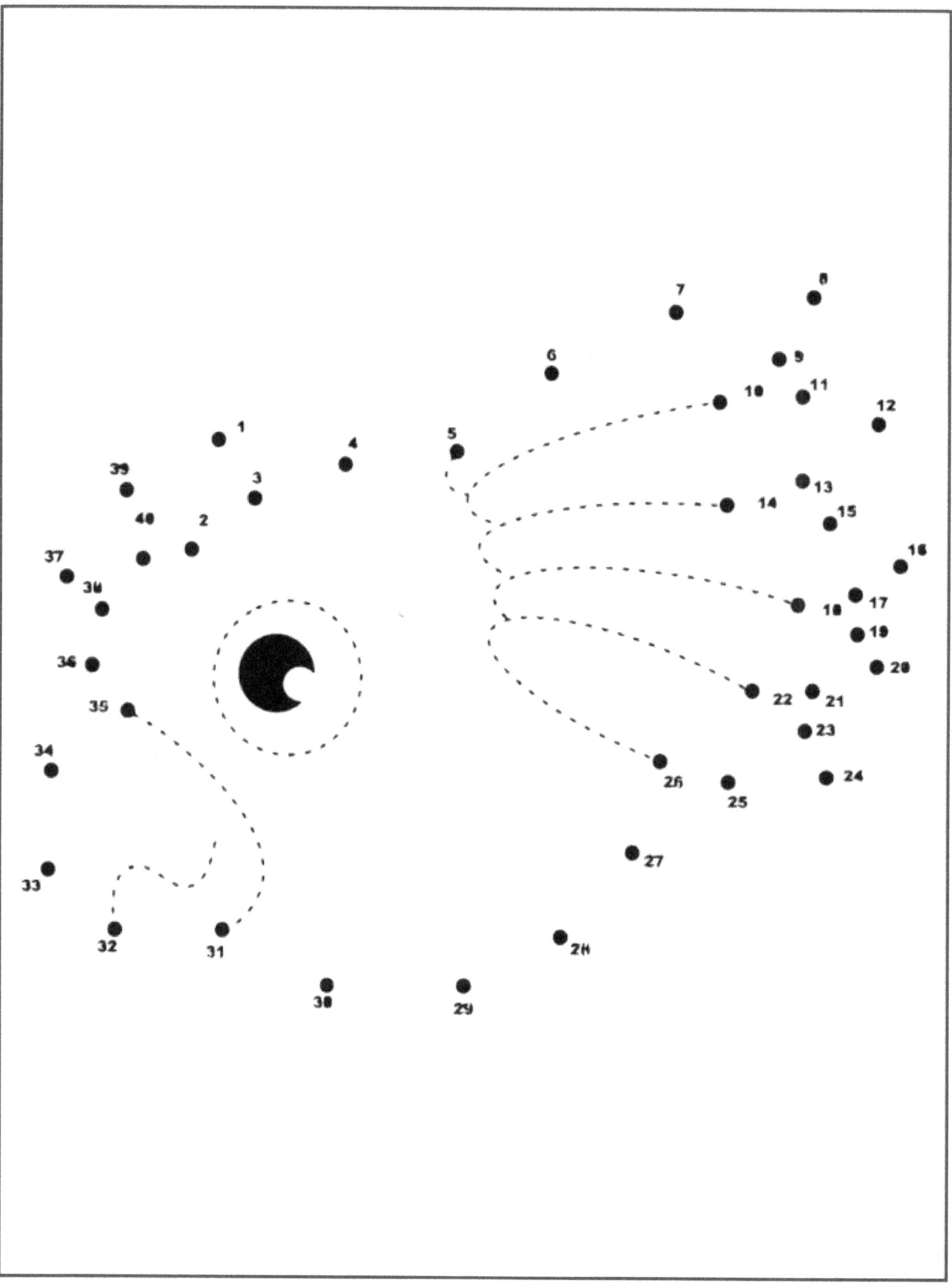

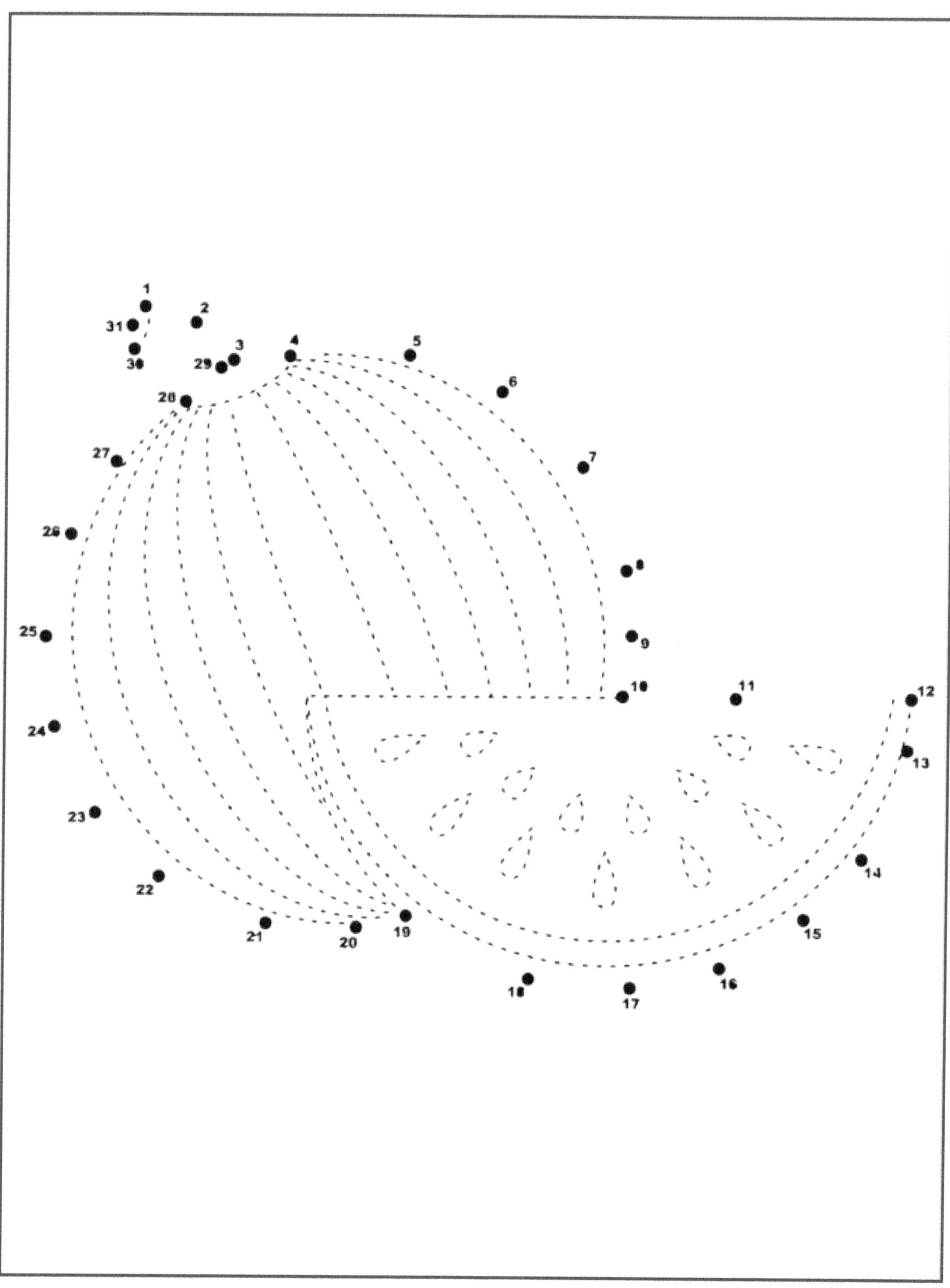

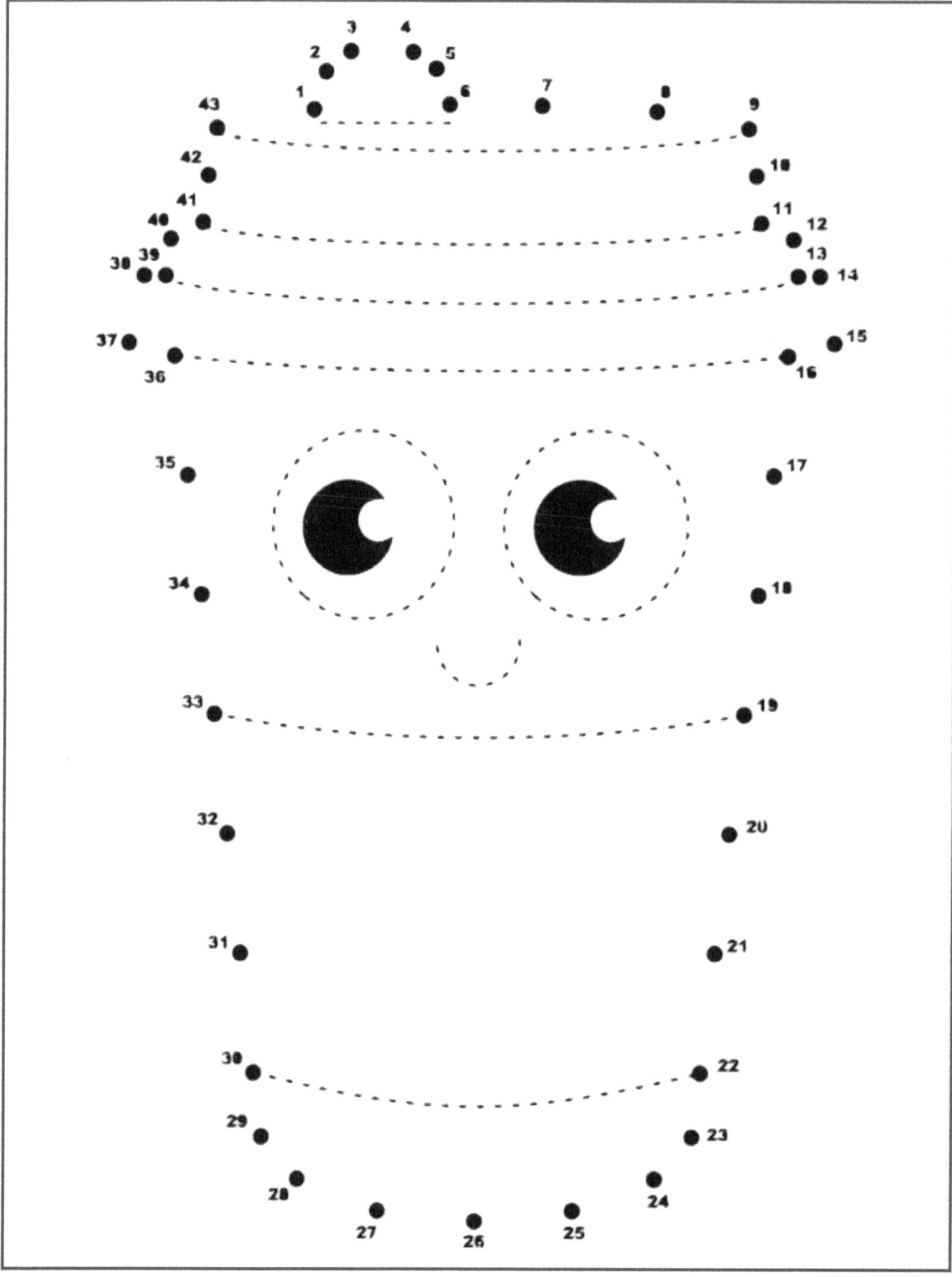

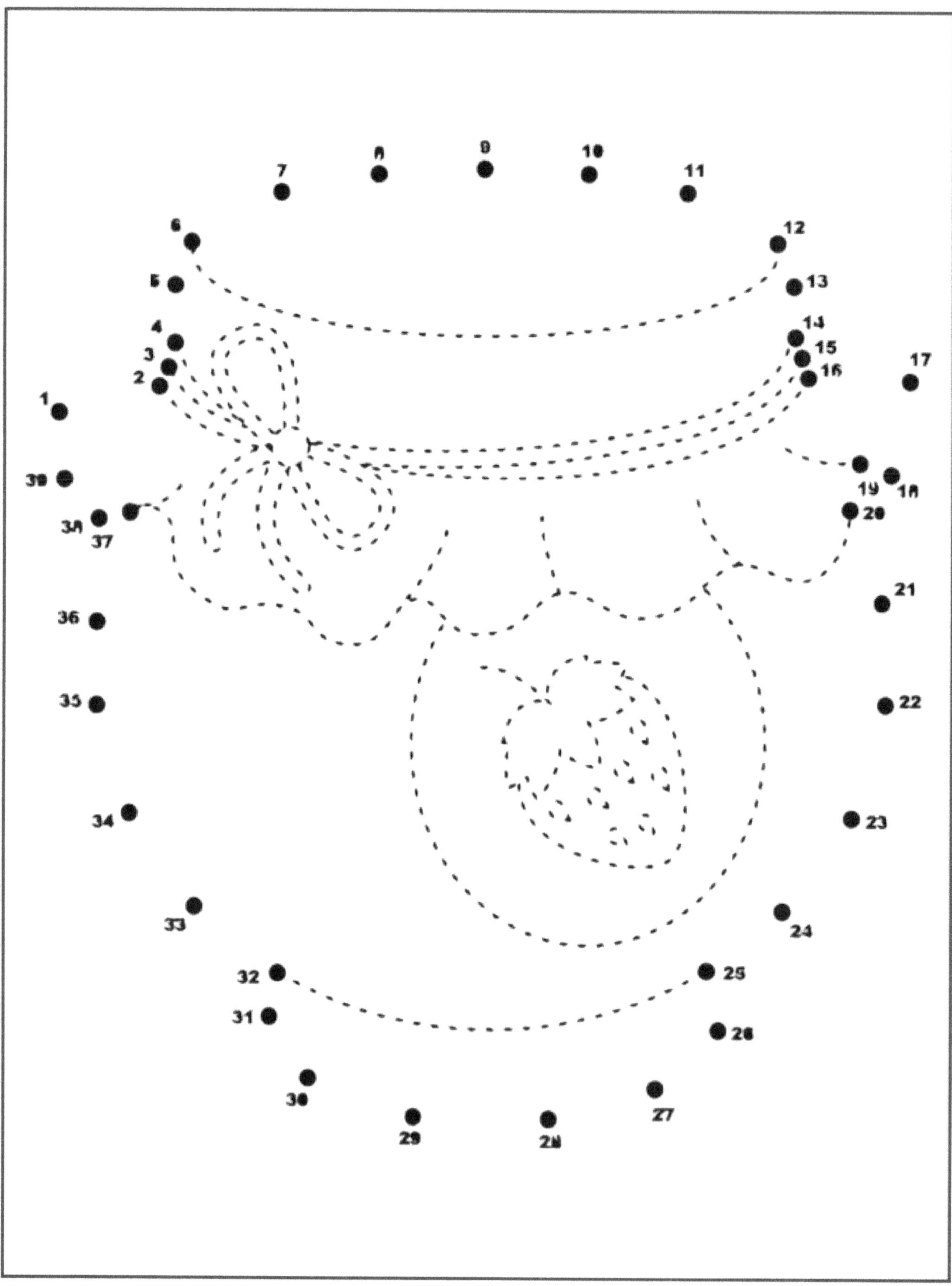

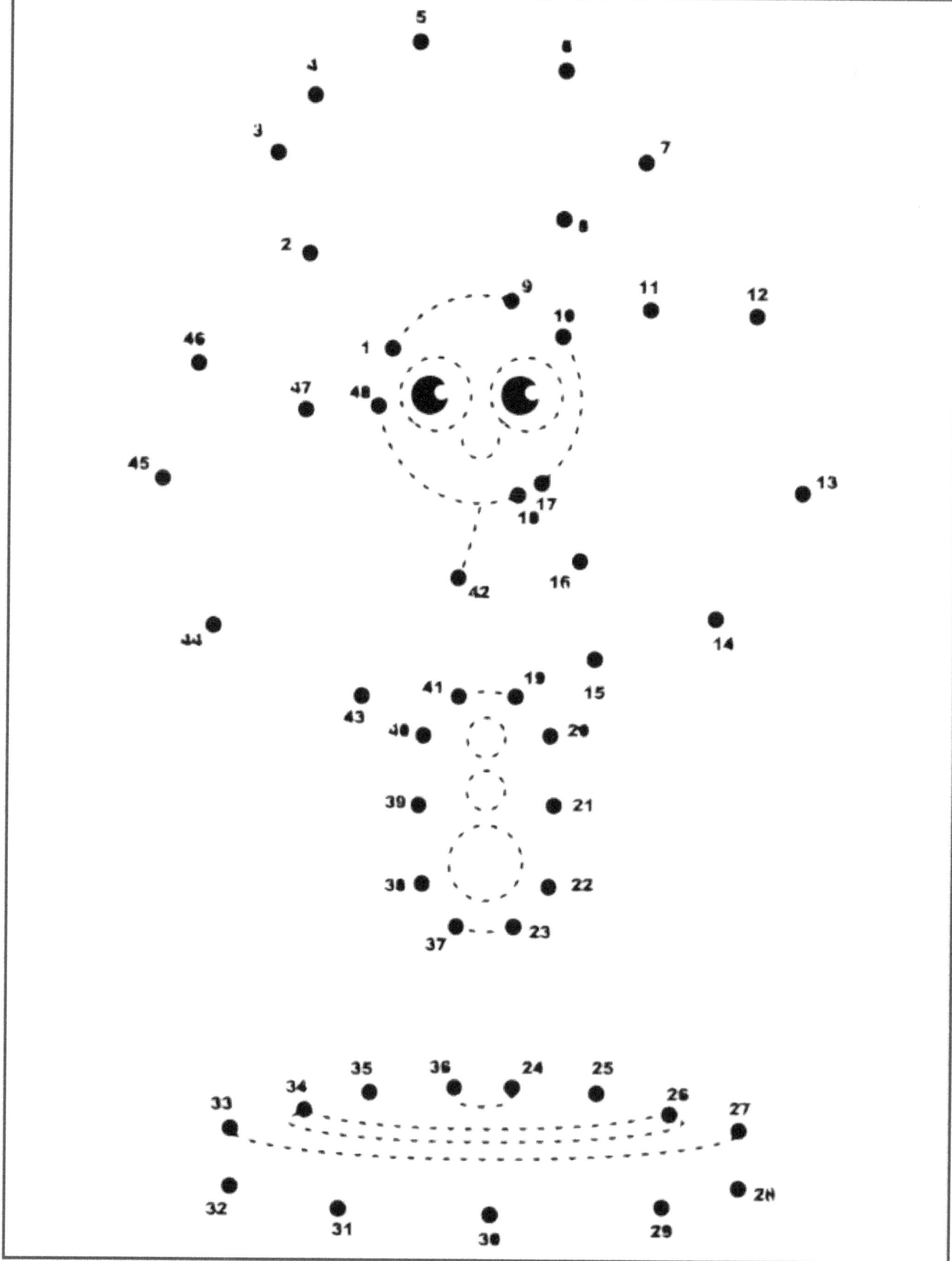

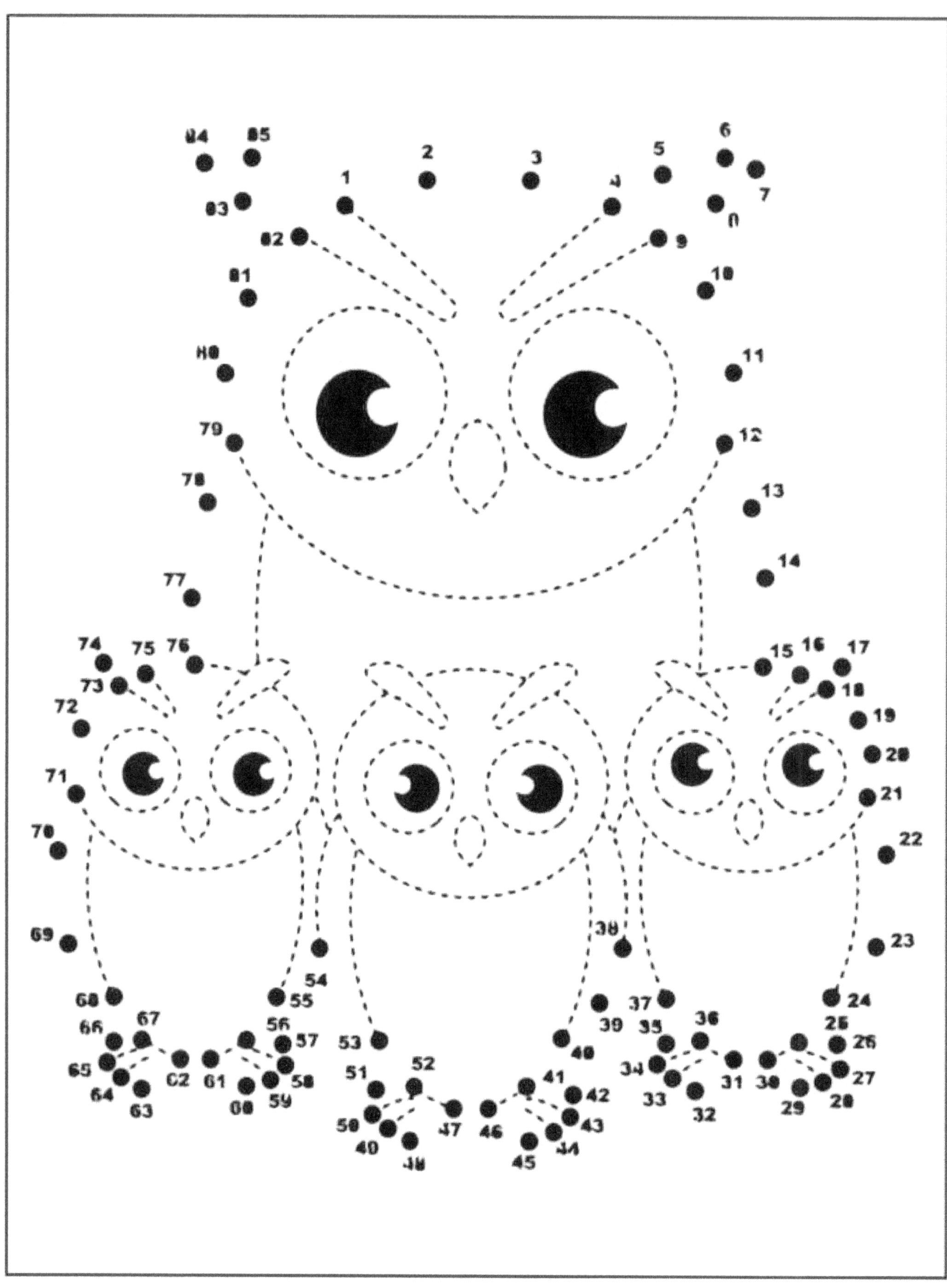

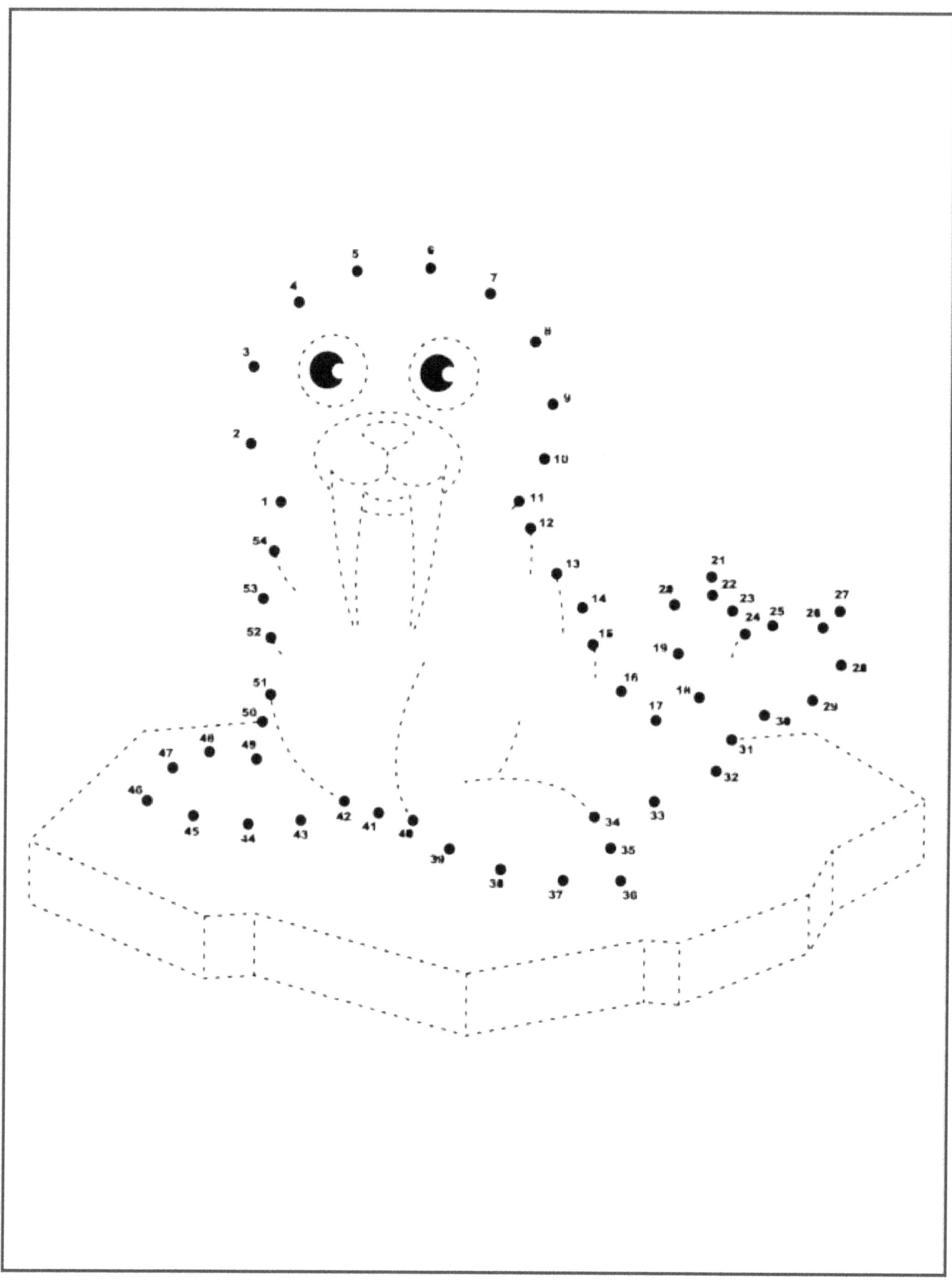

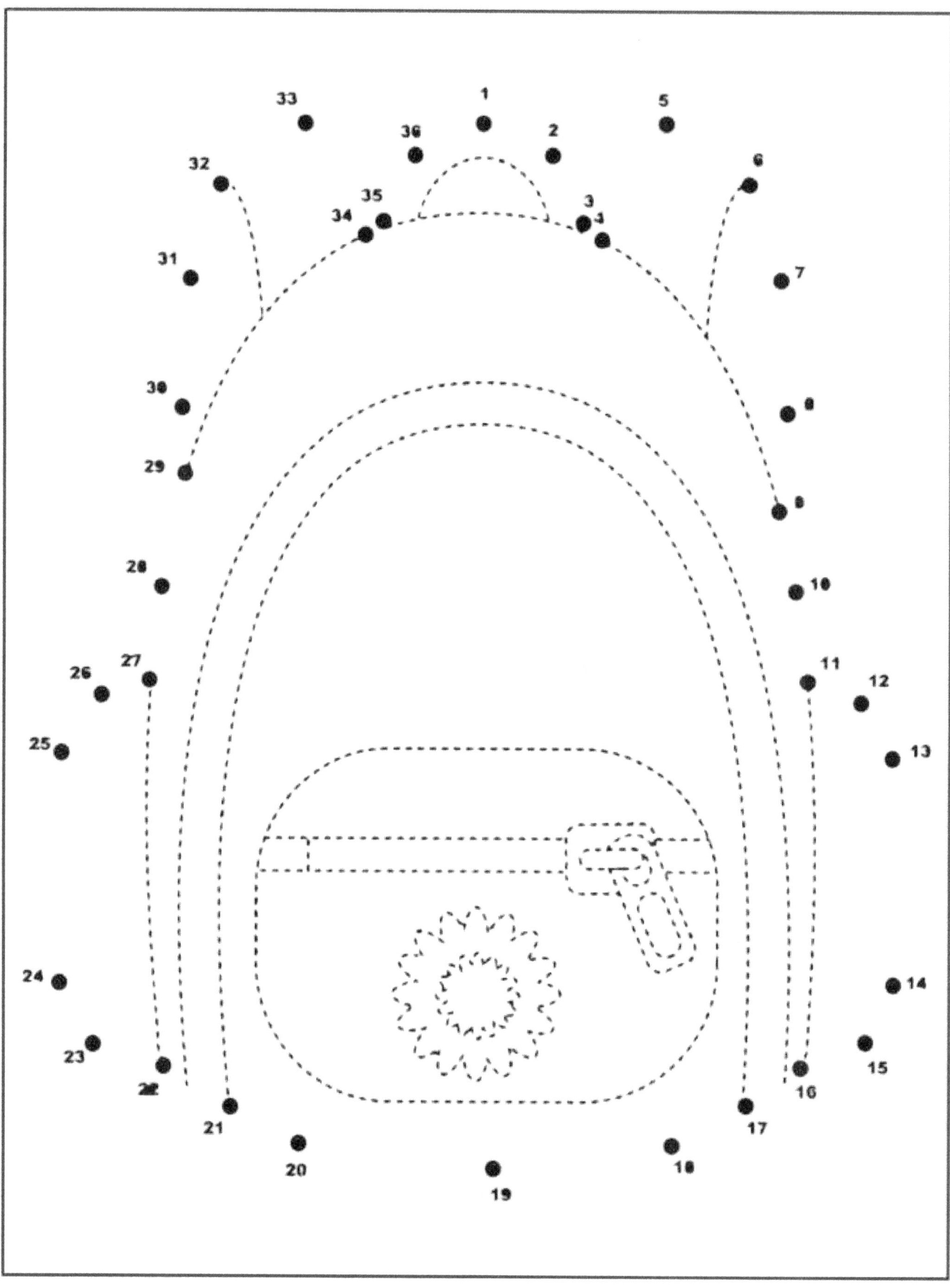

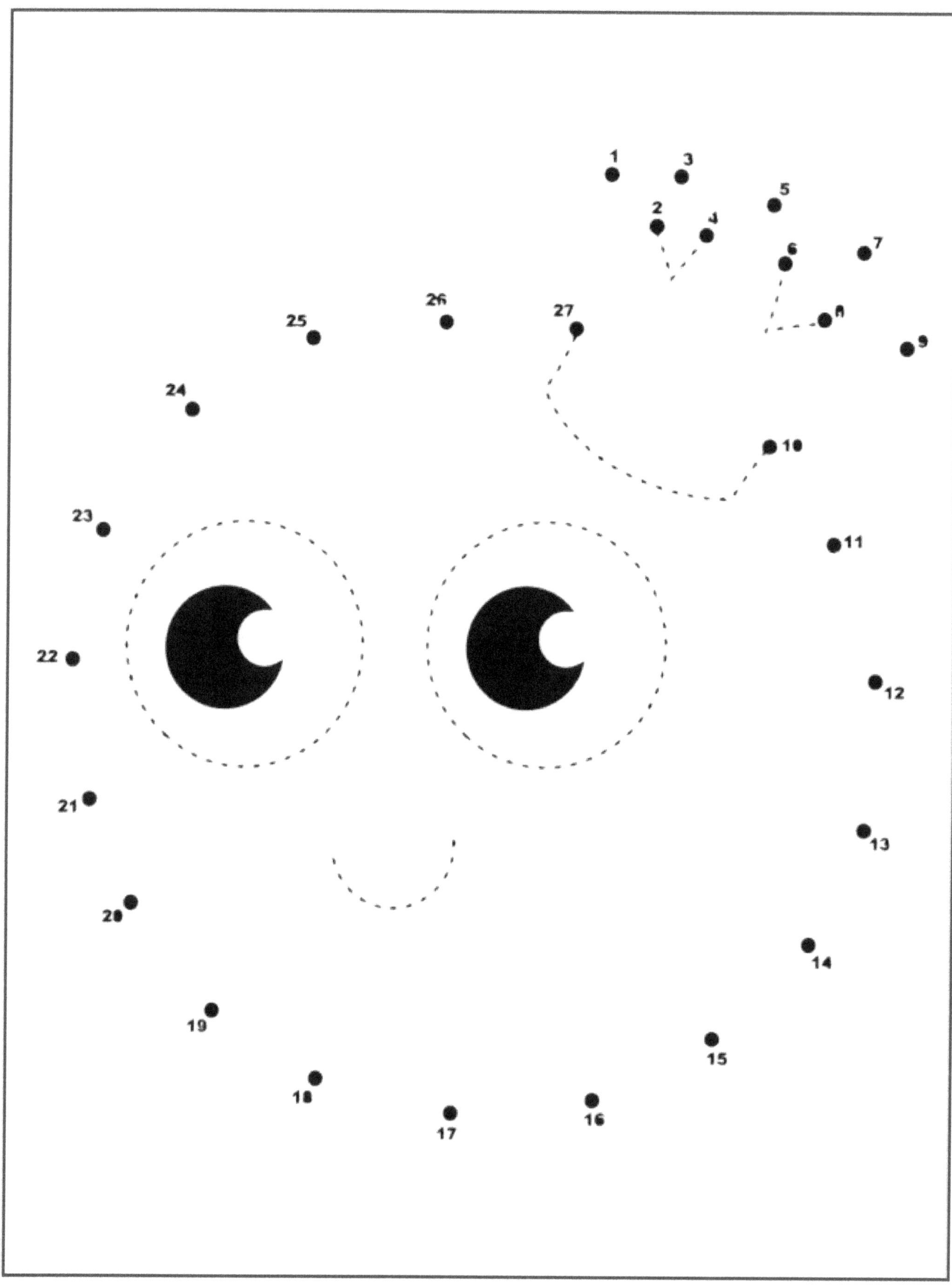

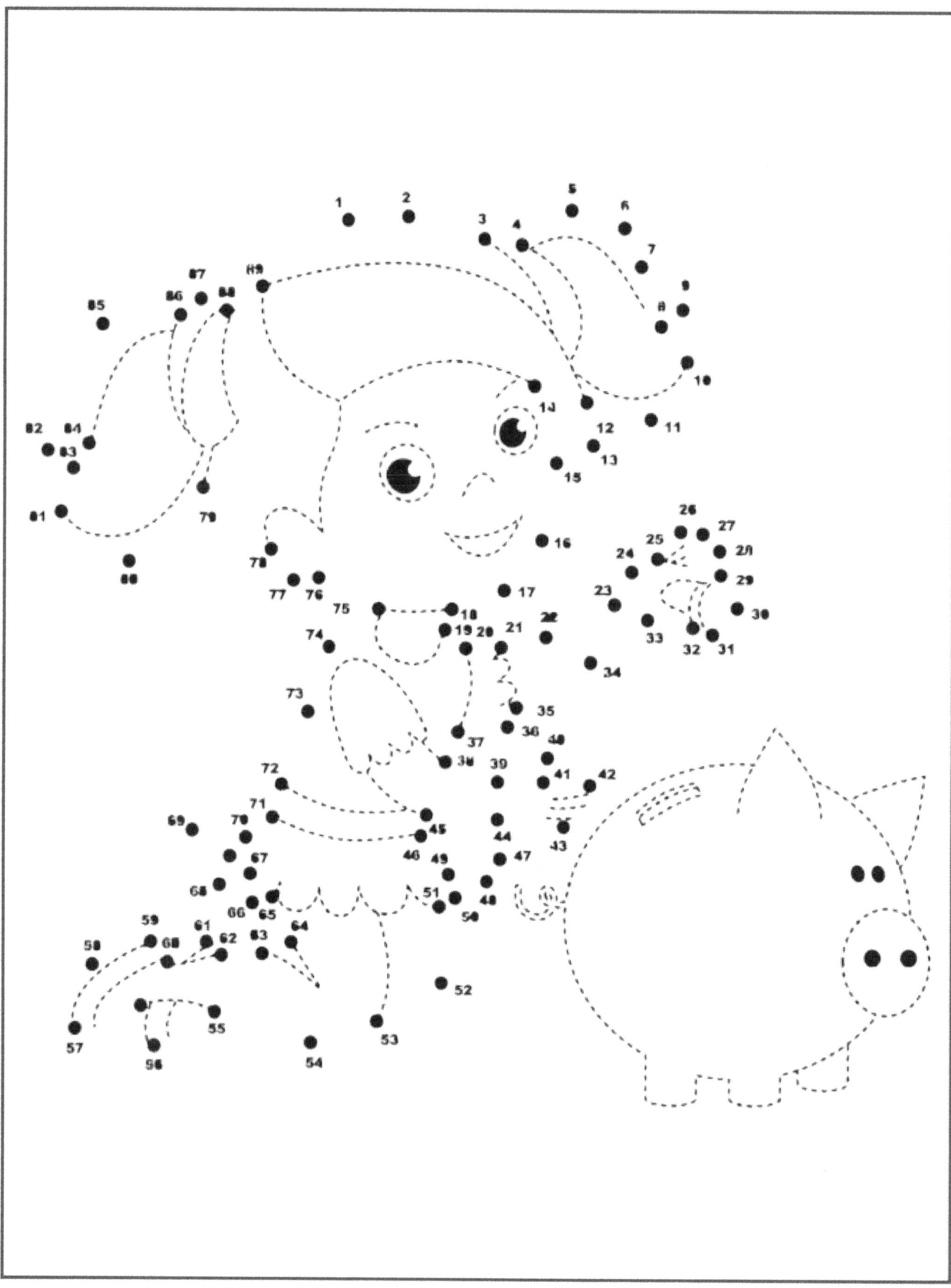

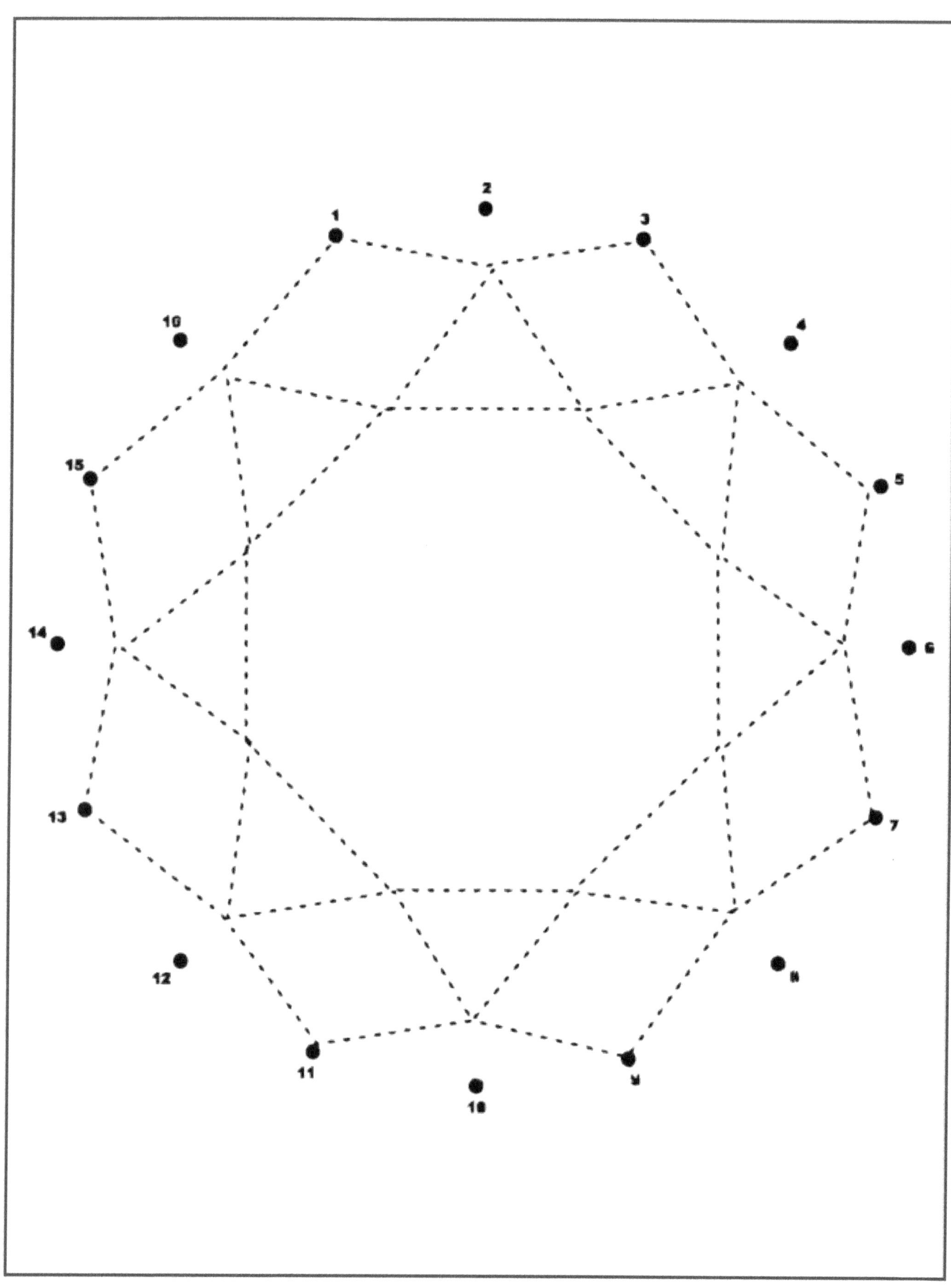

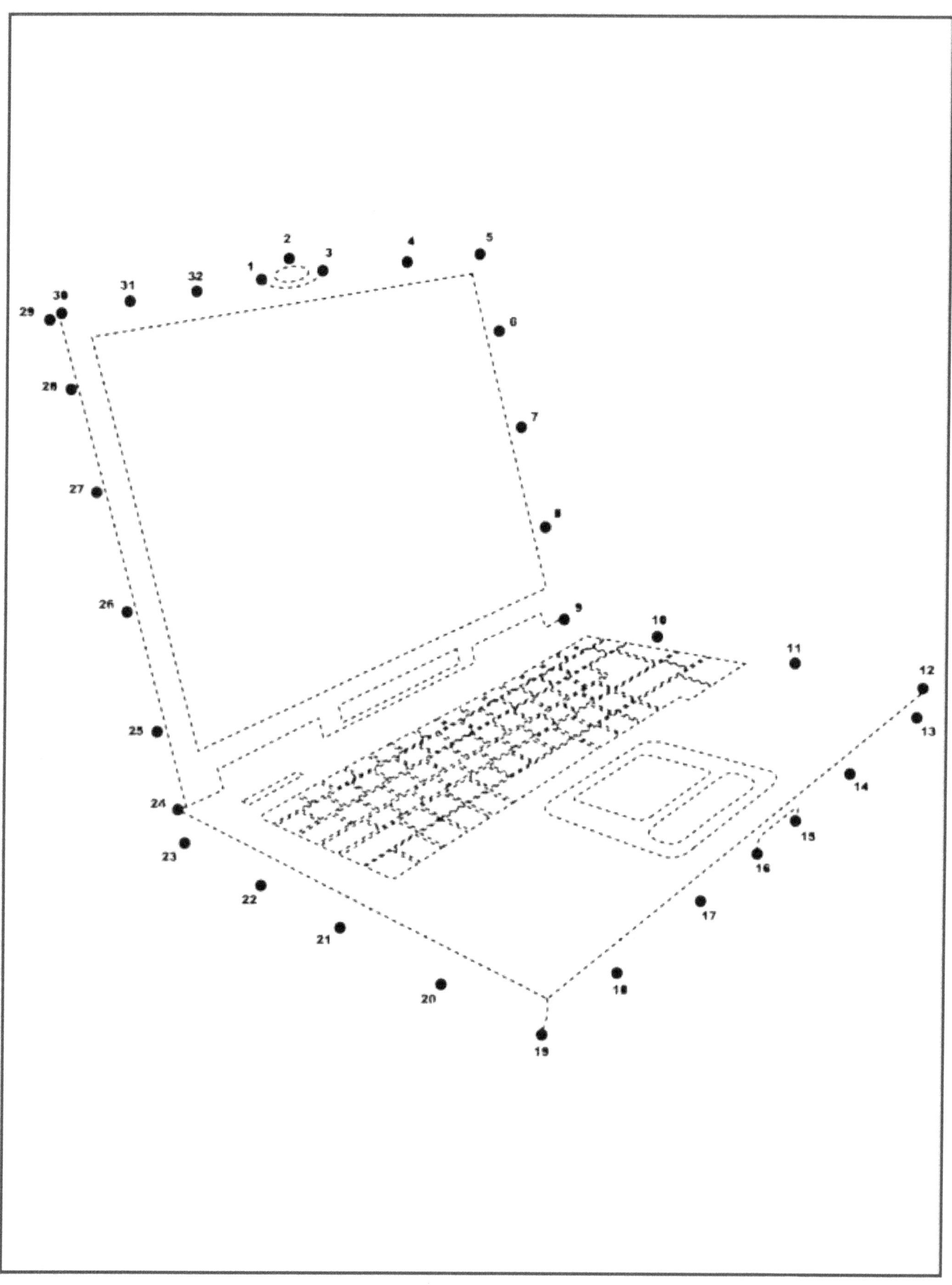

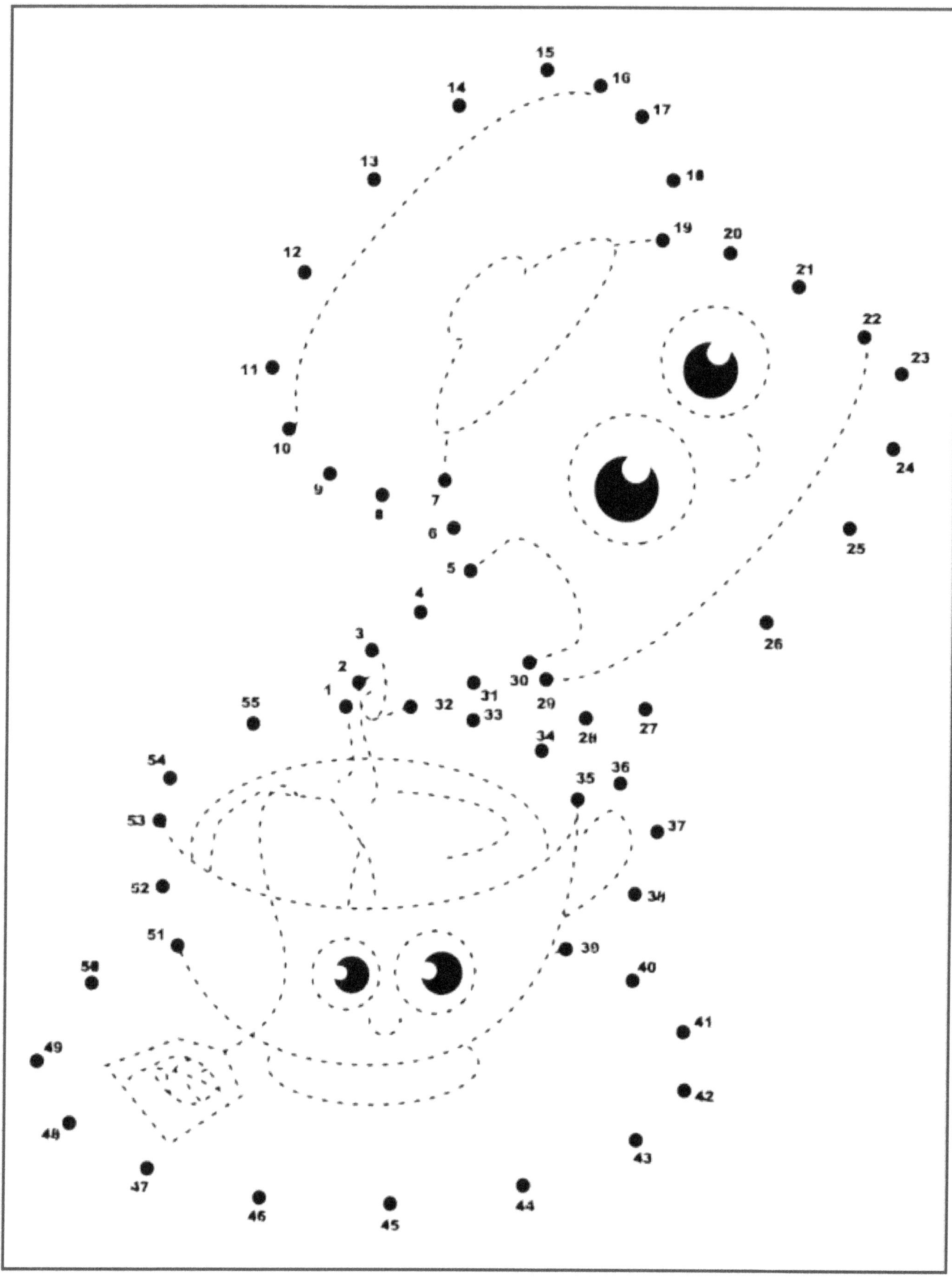

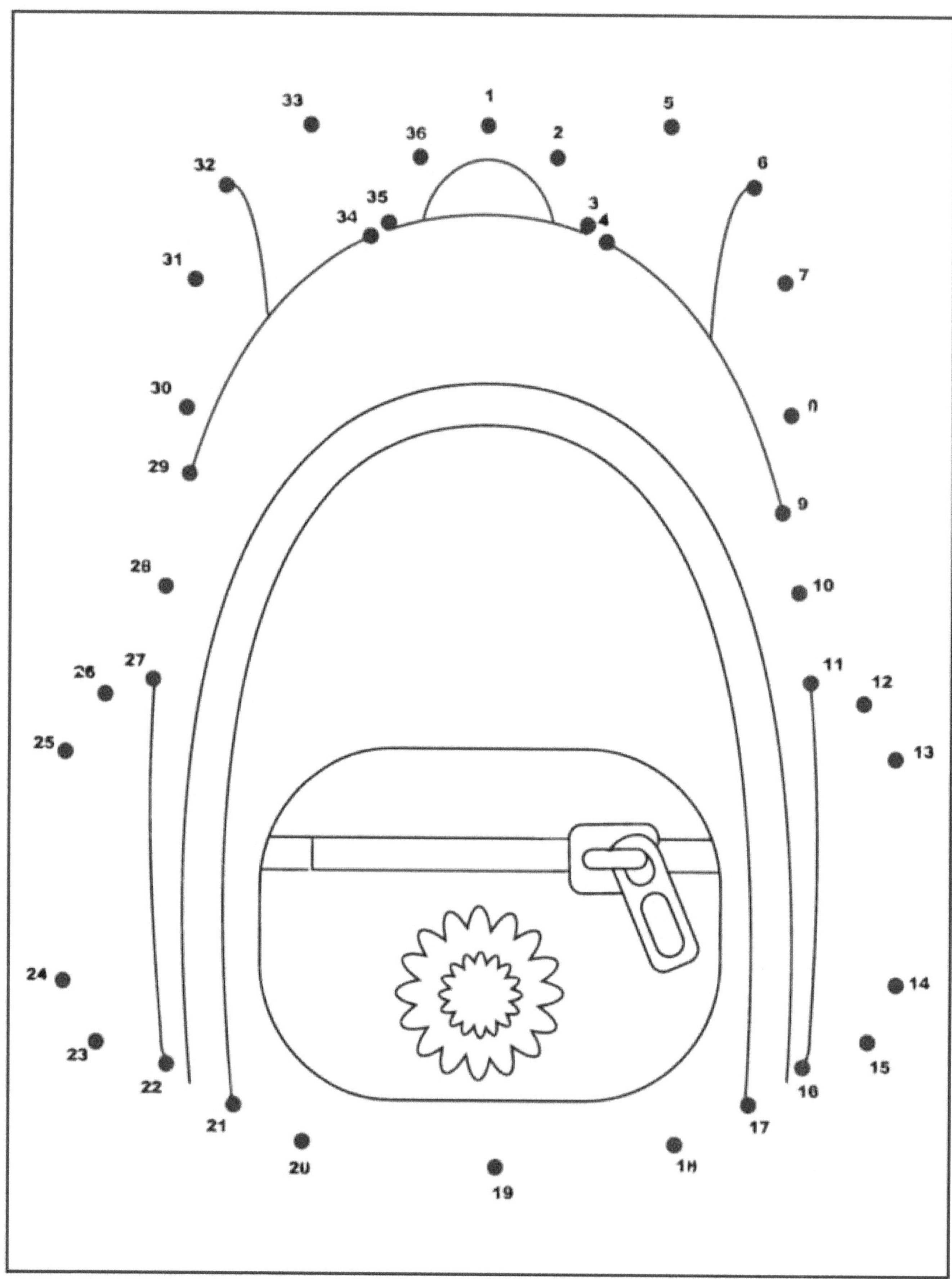

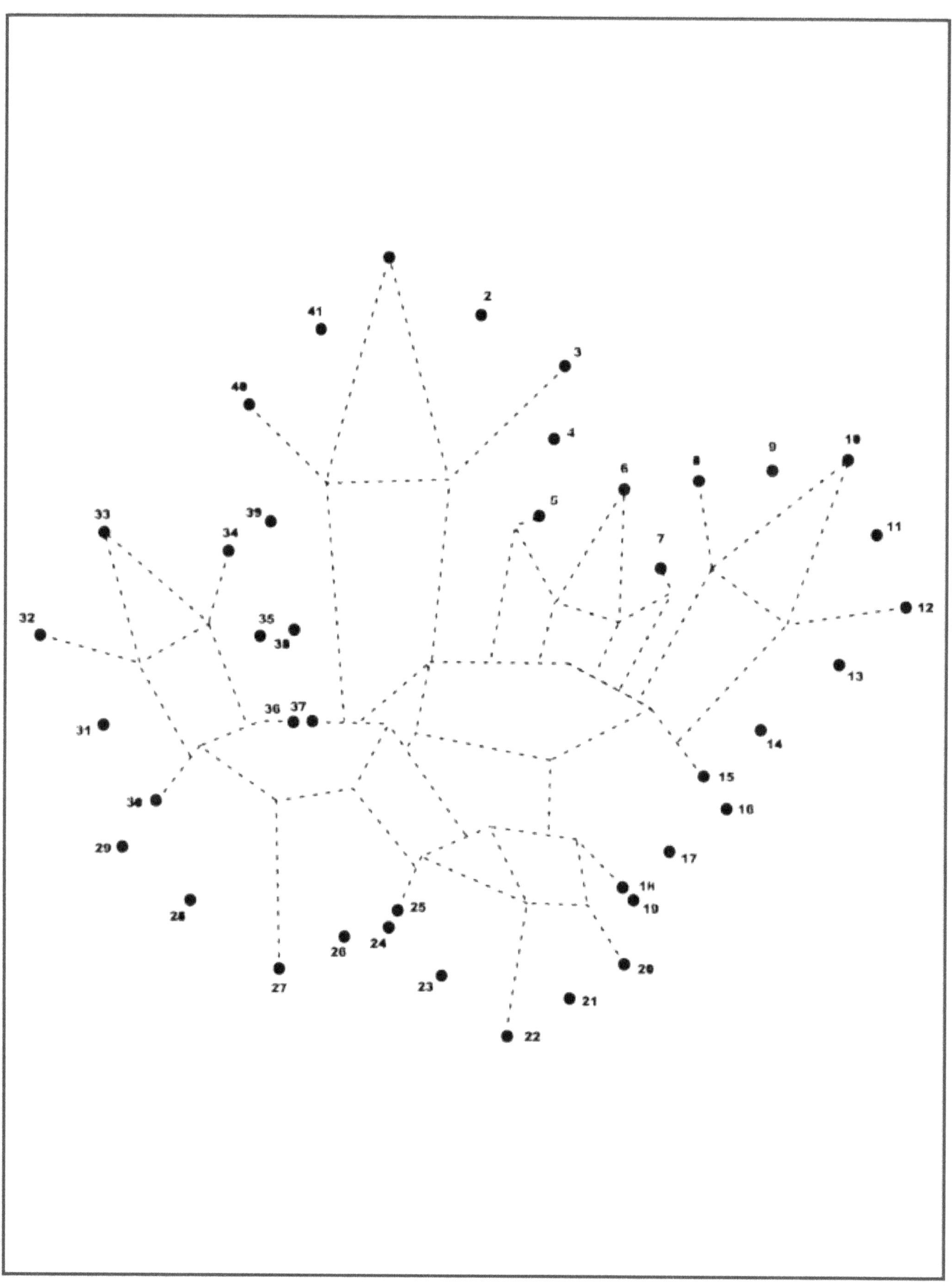

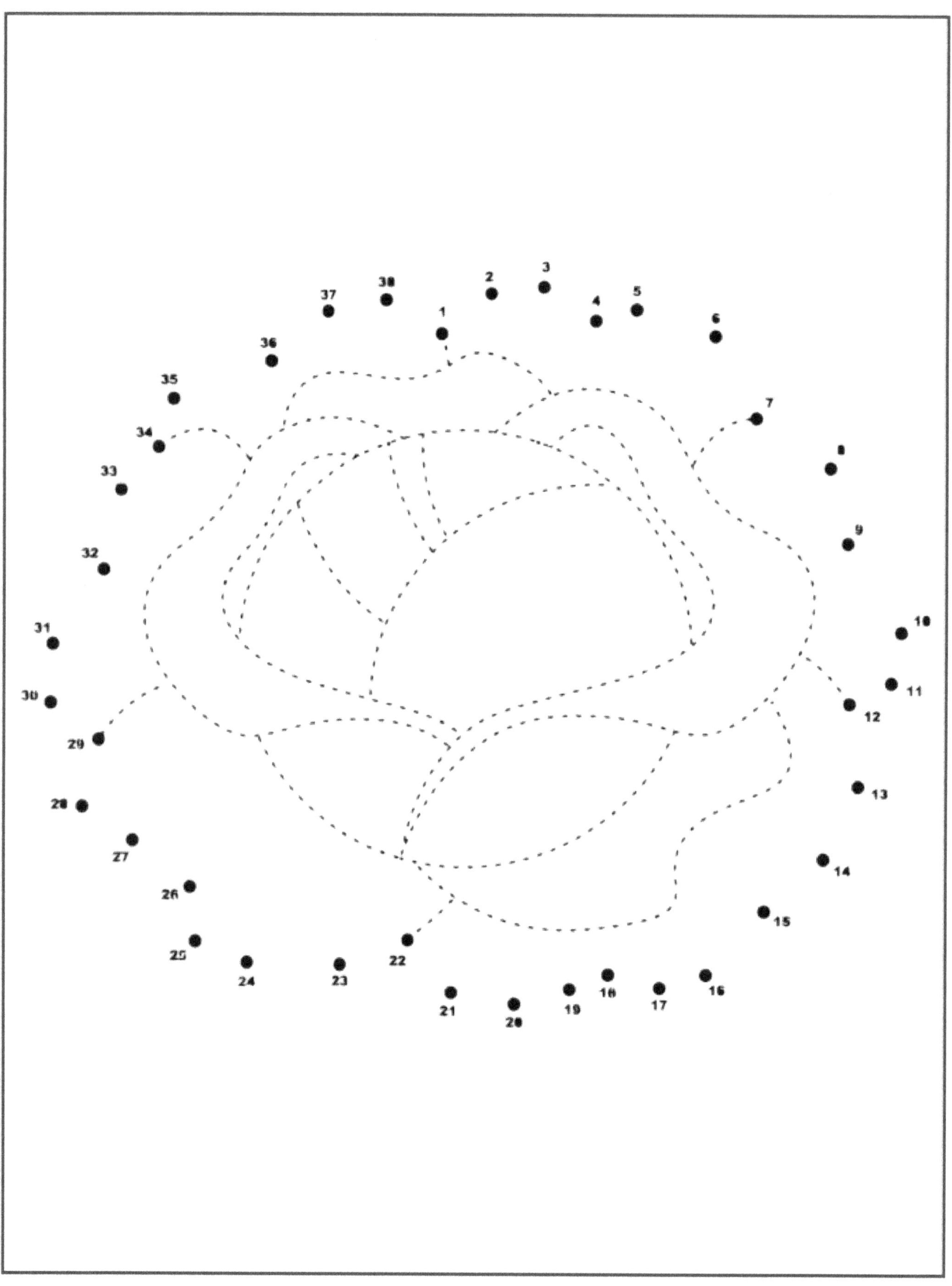

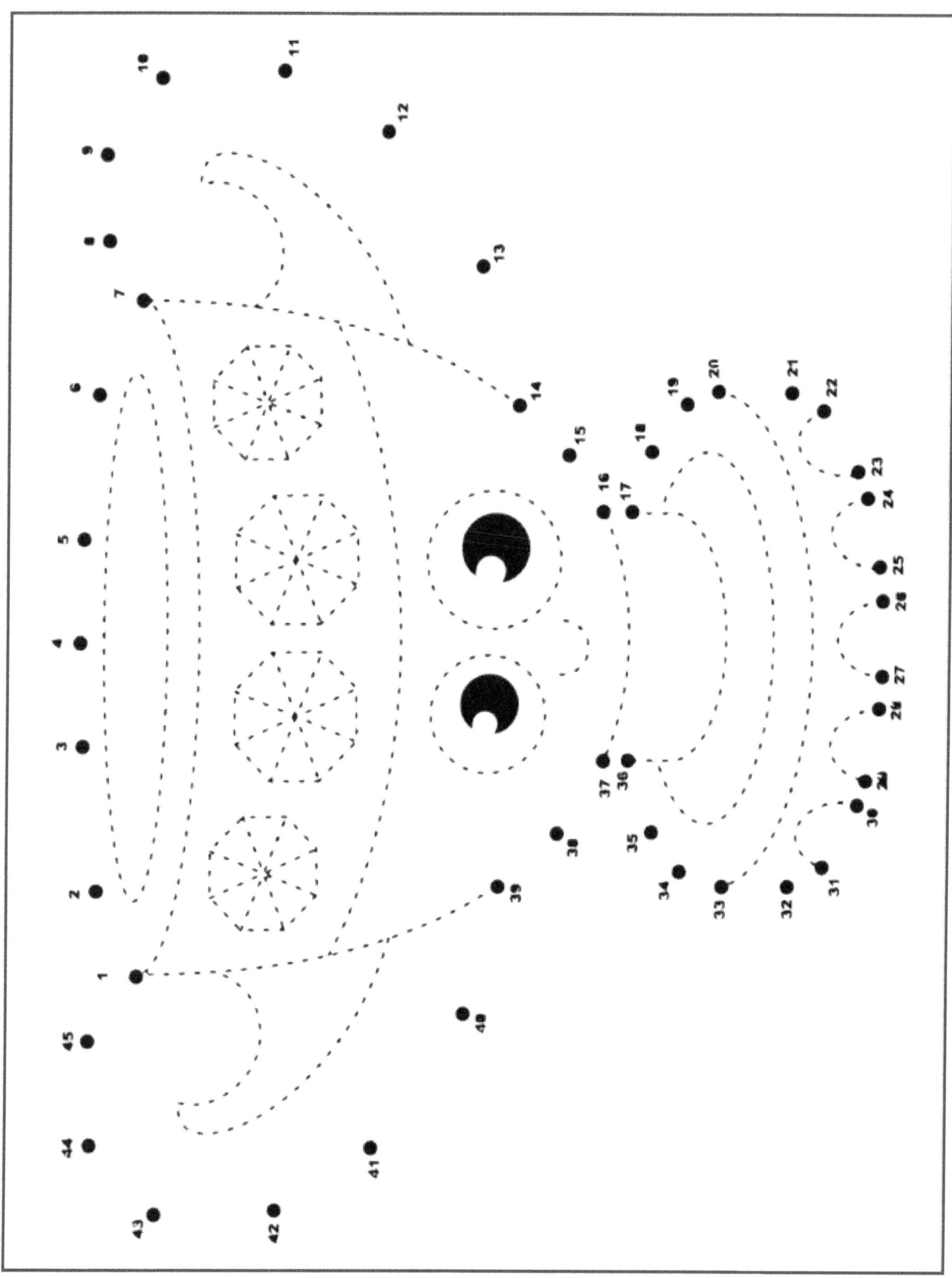

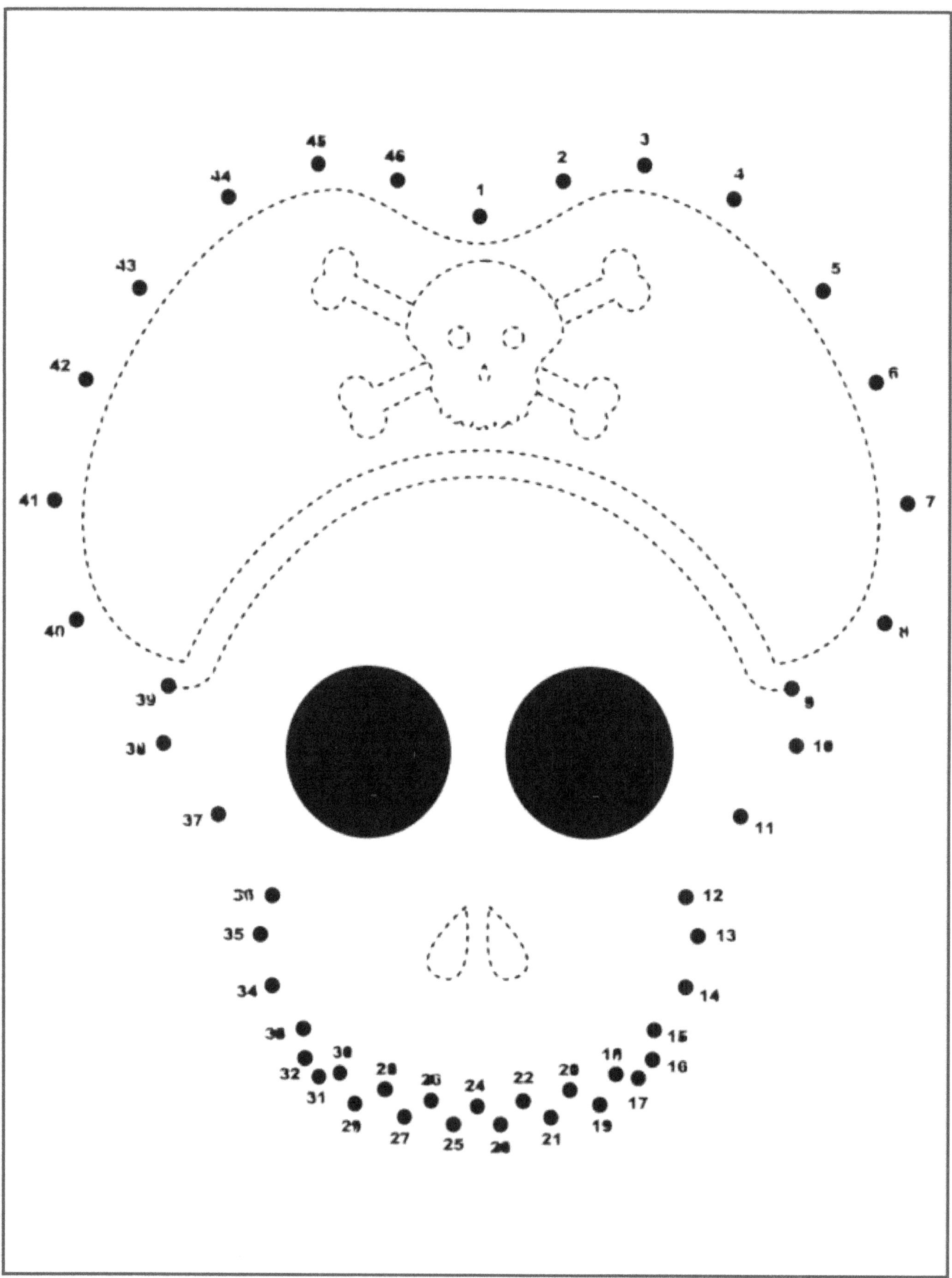

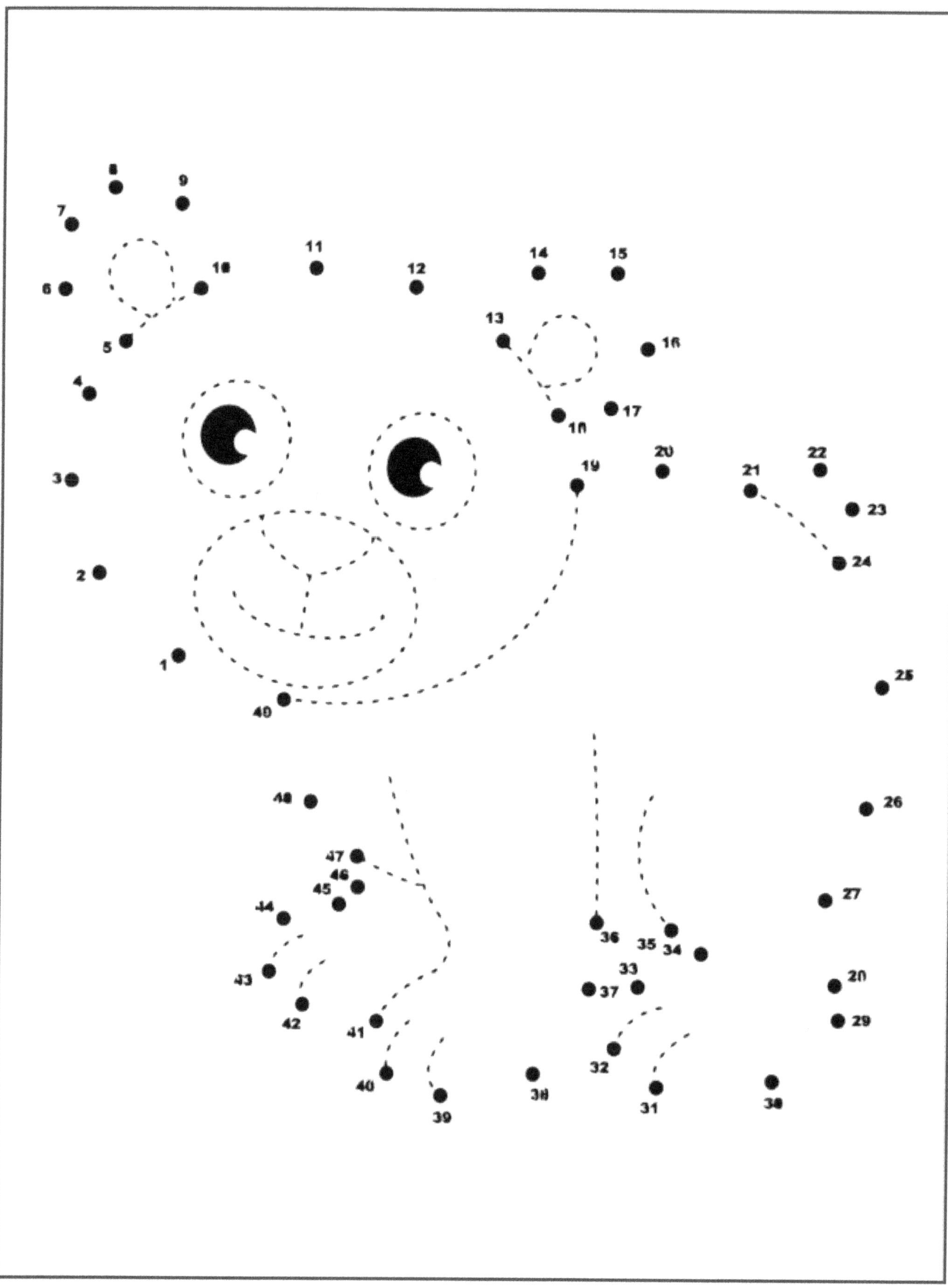

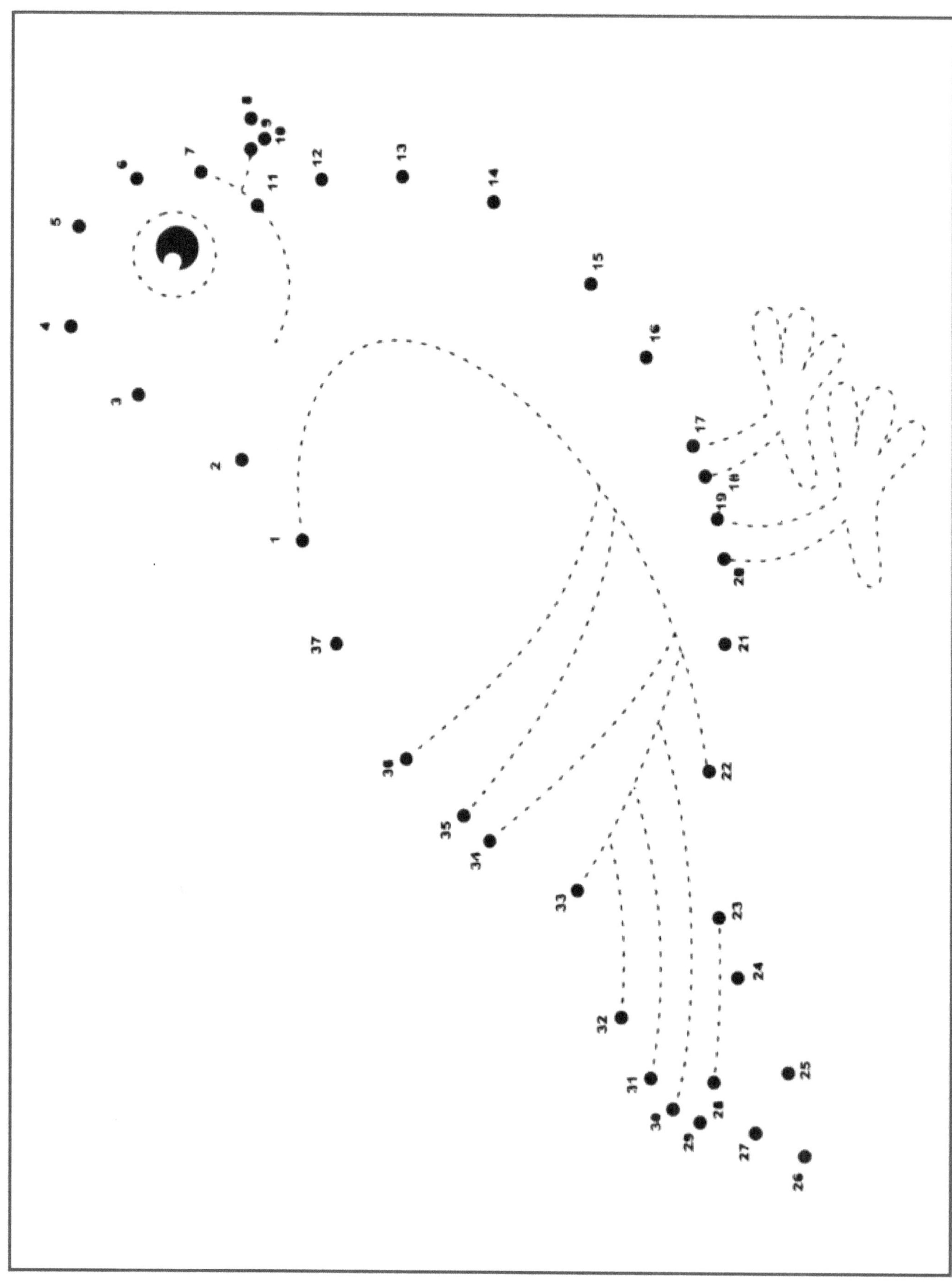

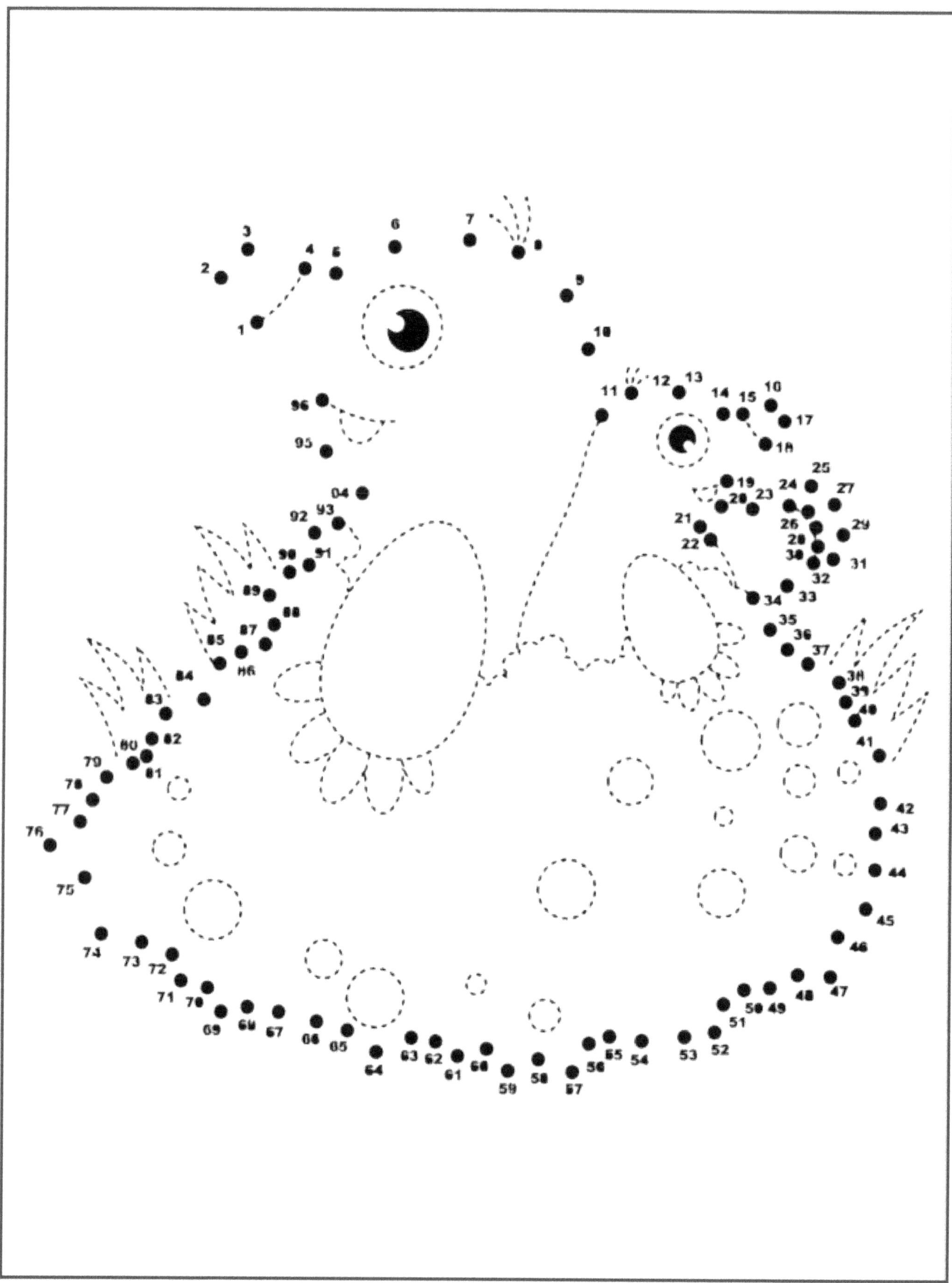

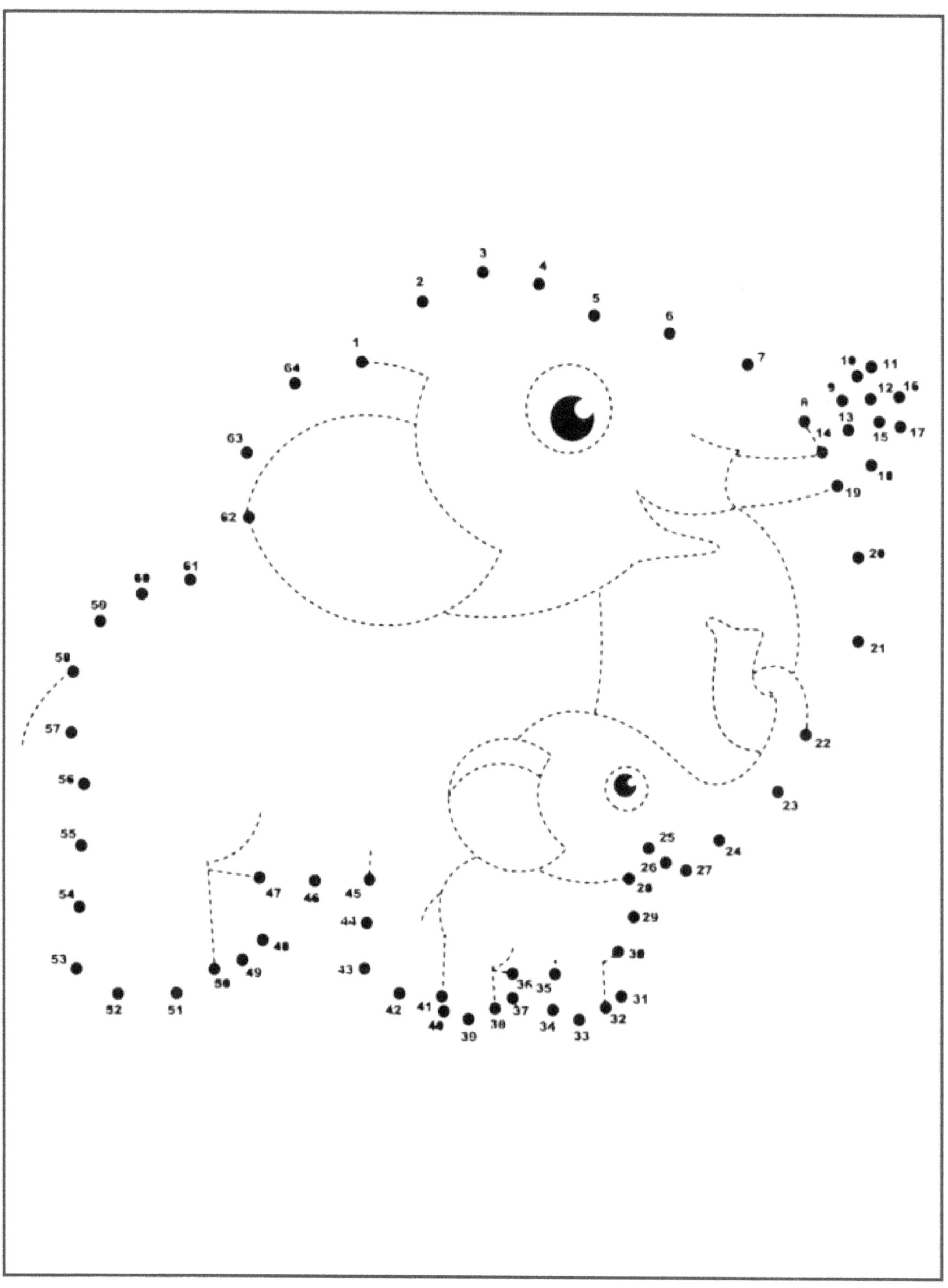